Initialization and Diversity in Optimization Algorithms

Mario Alberto Navarro Velázquez
Bernardo Morales Castañeda
Itzel Aranguren
Diego Oliva
Marco Perez-Cisneros
University of Guadalajara

CRC Press
Taylor & Francis Group
Boca Raton London New York

CRC Press is an imprint of the
Taylor & Francis Group, an **informa** business
A SCIENCE PUBLISHERS BOOK

First edition published 2026
by CRC Press
2385 NW Executive Center Drive, Suite 320, Boca Raton FL 33431

and by CRC Press
4 Park Square, Milton Park, Abingdon, Oxon, OX14 4RN

Library of Congress Cataloging-in-Publication Data (applied for)

ISBN: 978-1-032-69581-5 (hbk)
ISBN: 978-1-032-72413-3 (pbk)
ISBN: 978-1-032-72356-3 (ebk)

DOI: 10.1201/9781032723563

Typeset in Times New Roman
by Prime Publishing Services

Preface

Optimization lies at the core of scientific inquiry, engineering design, and real-world problem-solving. From minimizing costs and energy consumption to maximizing efficiency and system performance, optimization algorithms have transformed how we approach complex tasks. Among the plethora of nature-inspired metaheuristics, Differential Evolution (DE) and Particle Swarm Optimization (PSO) have emerged as two of the most powerful and widely used algorithms due to their simplicity, effectiveness, and adaptability across various domains. However, while much research has been devoted to the operational mechanics of these algorithms, relatively less focus has been placed on one of their most critical components: initialization and diversity.

This book, "Initialization and Diversity in Optimization Algorithms," is the result of years of research, experimentation, and deep reflection on the foundational mechanisms that underlie successful optimization. The central thesis of the work is straightforward yet profound: the initial distribution of candidate solutions and the maintenance of diversity during the search process have a significant influence on the performance and convergence behavior of population-based algorithms. While this principle is intuitively understood, its practical implications, nuanced strategies, and empirical validation remain underexplored in the literature, especially in the context of DE and PSO.

Our goal is to provide both a conceptual framework and a practical guide for researchers, graduate students, and practitioners seeking to enhance the effectiveness of their optimization routines. The book begins by revisiting the core structures of DE and PSO, highlighting how their search capabilities are intrinsically tied to the quality of their initial populations and the strategies they use to preserve or adapt diversity. Building on this foundation, we present a systematic study of various initialization techniques, including quasi-random sequences, chaotic maps, opposition-based learning, and stratified sampling. We then explore methods for measuring and controlling diversity throughout the optimization process, covering various metrics and indicators.

This work does not aim to replace existing studies on DE and PSO but rather to complement them by focusing on often-overlooked yet essential components. By shedding light on initialization and diversity, we aim to move the field toward more robust, reliable, and interpretable optimization techniques.

We hope this book inspires new research directions and empowers practitioners to reconsider the "first steps" of their optimization pipelines. After all, in any journey—whether through a search space or through knowledge itself—how you begin can shape everything that follows.

Contents

Chapter 1

Introduction to Swarm Optimization

Optimization plays a vital role in science, engineering, and decision-making, ensuring efficient solutions across various fields. Many real-world problems are highly complex, nonlinear, and high-dimensional, making traditional methods impractical. This challenge has led to the rise of metaheuristics—versatile algorithmic frameworks that offer robust, efficient solutions to complex optimization tasks.

This chapter explores metaheuristics emphasizing swarm intelligence, diversity's crucial role, and initialization strategies. It examines how these factors enhance the efficiency and effectiveness of optimization algorithms.

1.1 Metaheuristics: Beyond Exact Solutions

Metaheuristics are high-level problem-solving strategies designed for efficient search space exploration. Unlike exact methods that guarantee finding the global optimum, metaheuristics focus on obtaining near-optimal solutions within reasonable computational time. They are particularly useful for problems where:

- The search space is vast and complex: Enumerating all possible solutions is computationally infeasible.
- The objective function is non-linear and non-differentiable: Gradient-based optimization methods are not applicable.
- The problem is NP-hard or NP-complete: Finding an exact solution is computationally expensive.

DOI: 10.1201/9781032723563-1

Inspired by natural phenomena such as biological evolution, physical annealing, and social behavior, metaheuristics balance exploration (searching new regions) and exploitation (refining promising solutions) to optimize complex problems. The number of publications on metaheuristics continues to grow each year. Figure 1.1 shows the increase of publications related to metaheuristics from 1995 to 2024.

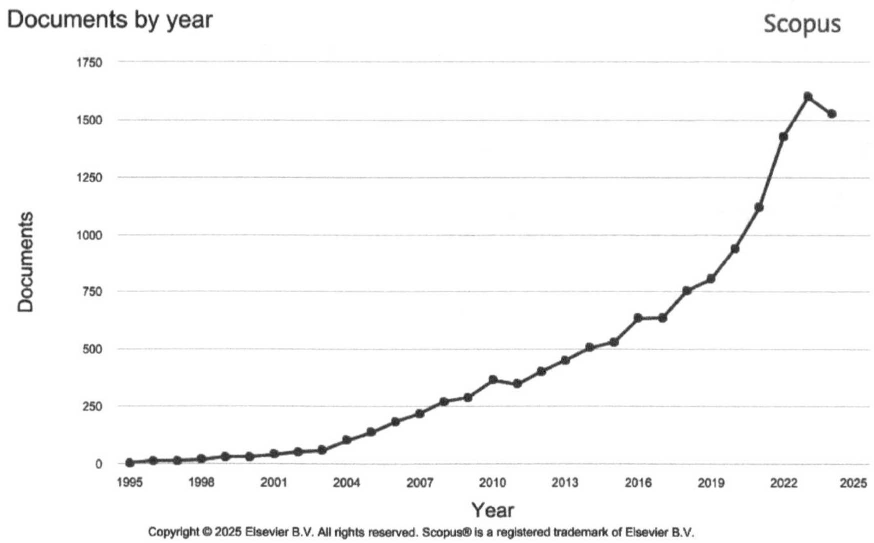

Figure 1.1: Yearly publications on metaheuristics (1995–2024).

A Metaheuristic Algorithm (MA) can operate with either a single solution or a population of candidate solutions. Throughout this book, the term "population" refers to a set of N candidate solutions, denoted as $Pop_p = [\mathbf{X}_1^p, \mathbf{X}_2^p, ..., \mathbf{X}_N^p]$. Initialization is the first step in any MA, where Pop_p is created and evaluated using the objective function. Through iterative evolution, operators modify solutions, forming Pop_{p+1} until a stopping criterion (often a predefined number of iterations) is met. The best solution, g^{best} is then extracted. The flowchart in Fig. 1.2 illustrates these steps.

Metaheuristic algorithms share key characteristics that define their behavior:

- **Approximation:** Focus on near-optimal rather than perfect solutions.
- **Stochasticity:** Use randomness to escape local optima and explore the search space.
- **Iteration:** Improve solutions progressively through repeated steps.
- **Flexibility:** Adaptable to various optimization challenges.

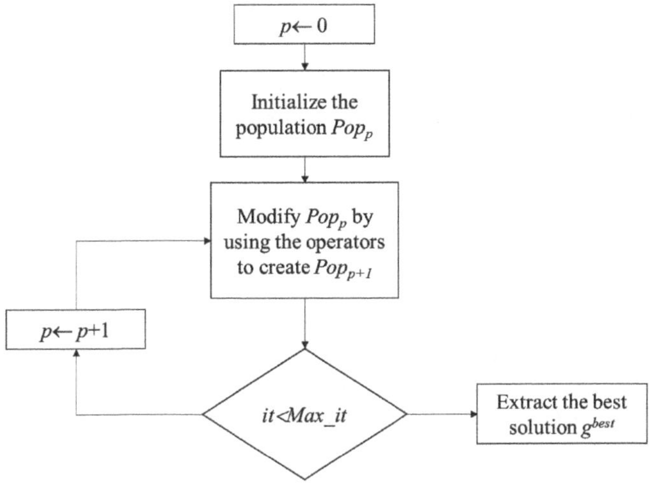

Figure 1.2: Standard metaheuristic algorithm flowchart.

Metaheuristics, inspired by various natural phenomena, are challenging to classify. Traditionally, they fall into three main categories:

- **Evolutionary Algorithms (EAs):** Modeled on biological evolution, they employ selection, crossover, and mutation to refine populations. Examples include Genetic Algorithms (GAs) [1], Evolutionary Programming (EP) [2], and Genetic Programming (GP) [3].
- **Swarm Intelligence (SI):** Based on the collective behavior, SI algorithms use agent interactions for optimization. Notable examples are Particle Swarm Optimization (PSO) [4], Ant Colony Optimization (ACO) [5], and Artificial Bee Colony (ABC) [6].
- **Trajectory-Based Methods:** These refine a single solution by guiding it through the search space. Common examples include Simulated Annealing (SA) [7] and Tabu Search (TS) [8].

1.2 Swarm Intelligence: Collective Wisdom

Swarm Intelligence (SI) is a subset of metaheuristics inspired by the collective behavior of social insects and animal groups. These algorithms use decentralized, self-organizing systems to tackle complex optimization problems. Despite varied inspirations, all SI methods follow key principles:

- **Decentralization:** Agents act independently, without central control.
- **Self-organization:** Global patterns emerge through local interactions.
- **Emergent behavior:** Intelligent solutions arise from simple individual actions.

Swarm Intelligence (SI) encompasses various algorithms inspired by natural behaviors. Classical approaches include:

- **Particle Swarm Optimization (PSO):** Based on flocking behavior, particles move through the search space, guided by personal and swarm-best positions.
- **Ant Colony Optimization (ACO):** Mimics ant foraging, using pheromone-based communication to find optimal paths.
- **Artificial Bee Colony (ABC):** Models honeybee foraging, dividing the swarm into employed onlooker, and scout bees, each with a distinct search role.

Some metaheuristics from other categories exhibit swarm-like behaviors. A notable example is Differential Evolution (DE) [9], which belongs to Evolutionary Algorithms (EAs) but shares characteristics with Swarm Intelligence (SI). DE operates on a population of candidate solutions, evolving them mutation, crossover, and selection to refine the search process.

Swarm intelligence algorithms are particularly well-suited for problems with high dimensionality, dynamic environments, and distributed search spaces. Their ability to adapt and self-organize makes them powerful tools for tackling complex optimization challenges.

1.3 Initialization in Metaheuristics

The initialization strategy of a metaheuristic algorithm significantly affects performance, influencing convergence speed, solution quality, and robustness [10]. It also plays a role in restarting solutions or avoiding suboptimal convergence. Common initialization methods include:

- **Random Initialization:** Generates solutions randomly in the search space but may perform poorly in complex landscapes [11].
- **Heuristic Initialization:** Uses domain-specific heuristics to improve initial solution quality and convergence speed [12].
- **Latin Hypercube Sampling (LHS):** Ensures a more even distribution across the search space than random initialization [11].
- **Opposition-based learning (OBL):** Creates an initial population and opposite population selecting the best solutions from both [13].
- **Clustering methods:** Grouping data points, initializing solutions within or near clusters to enhance search efficiency [14].

Choosing the right initialization strategy depends on the optimization problem and the metaheuristic algorithm. Testing different techniques helps identify the most effective approach. Due to its simplicity, random initialization is the most commonly used method.

1.4 The Role of Diversity in Metaheuristics

The success of a metaheuristic algorithm relies on maintaining diversity in the candidate solution population, preventing premature convergence to suboptimal solutions [15].

A lack of diversity limits exploration, reducing performance. Common techniques to enhance diversity include:

- **Distance-based measures:** Evaluate similarity by calculating solution distances.
- **Entropy-based measures:** Quantify diversity through population information content.
- **Crowding distance:** Ensures varied non-dominated solutions in Genetic Algorithms (GAs).
- **Neighborhood topologies:** Regulate information flow in Particle Swarm Optimization (PSO).
- **Mutation operators:** Adjust mutation rates in Genetic Algorithms to introduce variation.

Balancing exploration and exploitation is key to optimizing performance. Excessive exploration slows progress, while too much exploitation risks premature convergence.

1.5 Conclusions

This chapter introduces key concepts in optimization, metaheuristics, swarm intelligence, diversity, and initialization. The following chapters, will delve deeper into specific algorithms, advanced techniques for improving performance, and practical applications. We will explore mathematical foundations, diversity measurement, initialization methods, and parameter effects, equipping you with the knowledge to tackle complex optimization challenges and advance this field.

References

[1] D. E. Goldberg, *Genetic Algorithms*. Pearson Education India, 2013.

[2] D. B. Fogel, "An overview of evolutionary programming," in *Evolutionary Algorithms*. Springer, 1999, pp. 89–109.

[3] J. R. Koza, "Introduction to genetic programming," in *Proceedings of the 9th Annual Conference Companion on Genetic and Evolutionary Computation*, 2007, pp. 3323–3365.

[4] J. Kennedy and R. Eberhart, "Particle swarm optimization," in *Proceedings of ICNN'95-International Conference on Neural Networks*, vol. 4. IEEE, 1995, pp. 1942–1948.

[5] M. Dorigo and G. Di Caro, "Ant colony optimization: A new meta-heuristic," in *Proceedings of the 1999 Congress on Evolutionary Computation-CEC99 (Cat. No. 99TH8406)*, vol. 2. IEEE, 1999, pp. 1470–1477.

[6] D. Karaboga and B. Basturk, "A powerful and efficient algorithm for numerical function optimization: Artificial bee colony (abc) algorithm," *Journal of Global Optimization*, vol. 39, pp. 459–471, 2007.

[7] S. Kirkpatrick, C. D. Gelatt Jr, and M. P. Vecchi, "Optimization by simulated annealing," *Science*, vol. 220, no. 4598, pp. 671–680, 1983.

[8] F. Glover and M. Laguna, *Tabu Search*. Springer, 1998.

[9] R. Storn and K. Price, "Differential evolution–a simple and efficient heuristic for global optimization over continuous spaces," *Journal of Global Optimization*, vol. 11, pp. 341–359, 1997.

[10] B. Kazimipour, X. Li, and A. K. Qin, "A review of population initialization techniques for evolutionary algorithms," in *2014 IEEE Congress on Evolutionary Computation (CEC)*. IEEE, 2014, pp. 2585–2592.

[11] M. A. Navarro, D. Oliva, A. Ramos-Michel, B. Morales-Castaneda, D. Zaldívar, and A. Luque-Chang, "A review of the use of quasi-random number generators to initialize the population in meta-heuristic algorithms," *Archives of Computational Methods in Engineering*, vol. 29, no. 7, pp. 5149–5184, 2022.

[12] T. Tometzki and S. Engell, "Systematic initialization techniques for hybrid evolutionary algorithms for solving two-stage stochastic mixed-integer programs," *IEEE Transactions on Evolutionary Computation*, vol. 15, no. 2, pp. 196–214, 2010.

[13] S. Rahnamayan, H. R. Tizhoosh, and M. M. Salama, "A novel population initialization method for accelerating evolutionary algorithms," *Computers & Mathematics with Applications*, vol. 53, no. 10, pp. 1605–1614, 2007.

[14] I. Poikolainen, F. Neri, and F. Caraffini, "Cluster-based population initialization for differential evolution frameworks," *Information Sciences*, vol. 297, pp. 216–235, 2015.

[15] D. Sudholt, "The benefits of population diversity in evolutionary algorithms: a survey of rigorous runtime analyses," *Theory of Evolutionary Computation: Recent Developments in Discrete Optimization*, pp. 359–404, 2019.

Chapter 2

Two Classical Metaheuristics: Differential Evolution and Particle Swarm Optimization

2.1 Introduction

Metaheuristic algorithms have played a vital role in optimization since their inception [1]. Using stochastic processes, they generate or refine heuristics to address complex problems. They are especially useful for solving intricate nonlinear optimization challenges, where finding an optimal solution is highly difficult [2], [3].

Most metaheuristic algorithms use a population of search agents following specific behaviors to explore feasible solutions. This approach promotes interaction and diversity, enhancing search efficiency [4], [5].

Differential Evolution (DE) [6] and Particle Swarm Optimization (PSO) [7] are widely applied in fields, like engineering and financial optimization. These methods are robust, efficiently handling complex problems while remaining easy to implement with minimal parameter tuning [8], [9].

DOI: 10.1201/9781032723563-2

This chapter introduces Differential Evolution and Particle Swarm Optimization, covering their theories, computational frameworks, and mechanisms. Readers will gain insight into their advantages and limitations, enabling effective application to real-world optimization challenges.

2.2 Differential Evolution

Differential Evolution (DE), introduced by Storn and Price in 1997 [6], is a population-based metaheuristic known for its simplicity, efficiency, and robustness in optimization. Inspired by evolutionary principles, it enhances candidate solutions iteratively through mutation, crossover, and selection to refine results.

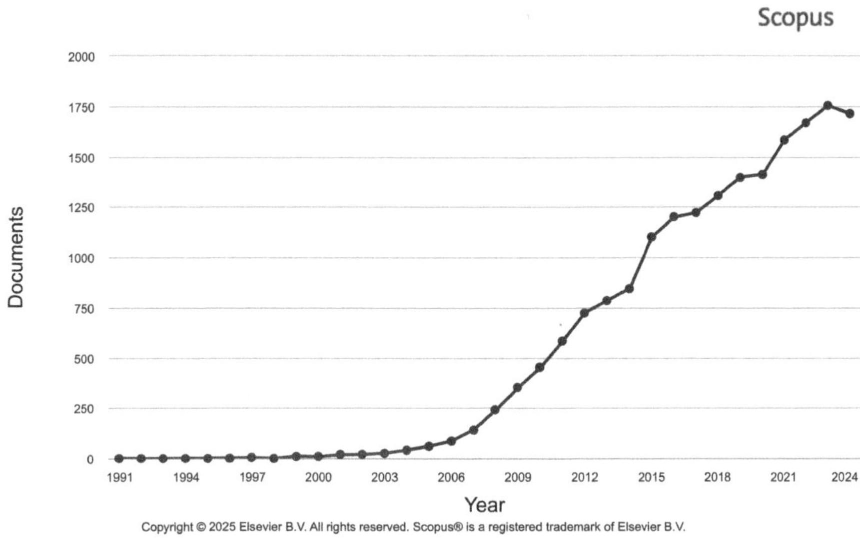

Figure 2.1: Documents by year related to DE approaches.

Differential Evolution (DE) follows an iterative process, starting with a randomly generated population of search agents spread within the search space boundaries. This population evolves through mechanisms inspired by natural evolution:

- Initialization: Generate an initial population of np individuals, each represented as a D-dimensional vector.
- Mutation: Modify population vectors to create mutant vectors.
- Crossover: Blend mutant and target vectors to form trial vectors.
- Selection: Evaluate trial vectors and retain the best ones for the next generation.

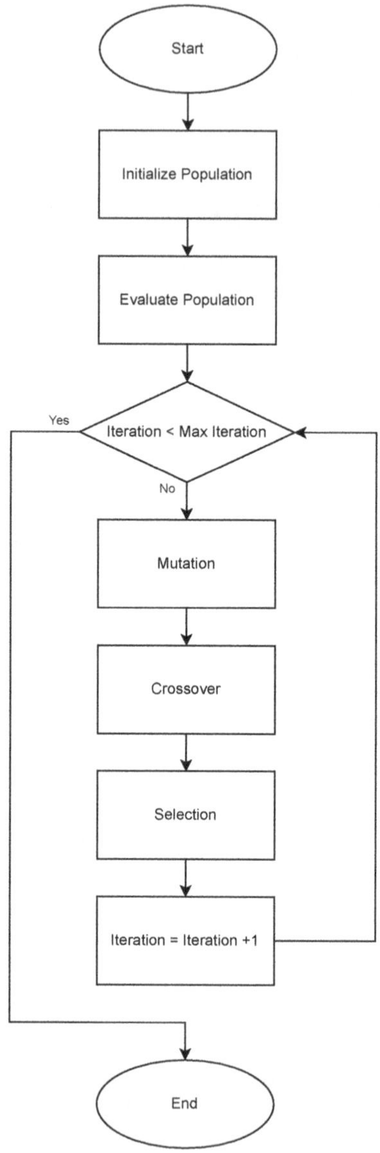

Figure 2.2: Flow diagram of the Differential Evolution Algorithm.

2.2.1 *Mutation*

The mutation process creates modified solutions for each individual in the population. Given $X = X_1, X_2, \ldots, X_np$, where np is the population size, mutation follows the formula:

$$M_i = X_c + F(X_a - X_b) \tag{2.1}$$

Table 2.1: Mutation strategies.

Strategy Name	Definition of Mutation Strategy
Rand/2	$M_i = X_f + F(X_a - X_b) + F(X_c - X_d)$
Best/1	$M_i = Xbest + F(X_a - X_b)$
Best/2	$M_i = Xbest + F(X_a - X_b) + F(X_c - X_d)$
Current to best/1	$M_i = X_i + F(Xbest - X_a) + F(X_b - X_c)$
Current to personal best/1	$M_i = X_i + F(X_p - X_i) + F(X_a - X_b)$

Here, a, b and c are randomly selected population indexes chosen at random, while F controls the magnitude of the differential variation $(X_a - X_b)$. Equation 2.1 represents the basic "rand/1" mutation strategy, though many variants exist.

In these equations, a, b, c, d, are randomly chosen indexes, $Xbest$ is the best individual in the population, and p is a randomly selected index from the top P individuals.

2.2.2 *Crossover*

The crossover process merges each solution with its "mutated" counterpart. Given $X_i = x_1, x_2, \ldots, x_n$, and $M_i = m_1, m_2, \ldots, m_n$, combined solution $C_i = c_1, c_2, \ldots, c_n$ is formed as:

$$c_j = \begin{cases} x_j, & \text{if } rand(0,1) \leq C_r \\ m_j, & \text{otherwise} \end{cases} \quad for j = 1, \ldots, n \tag{2.2}$$

Here, C_r is the crossover rate, determining whether elements come from the original or the mutated solution.

2.2.3 *Selection*

In the selection process, the combined solution is evaluated against the original based on fitness. The better solution advances to the next generation, continuing the iterative cycle:

$$X_i = \begin{cases} C_i, & \text{if } fitness(C_i) \leq fitness(X_i) \\ X_i, & \text{otherwise} \end{cases} \tag{2.3}$$

2.2.4 *Conclusion and Applications*

Differential Evolution (DE) is widely recognized for its effectiveness in solving non-linear, non-differentiable, and multi-modal optimization problems. Its simplicity and robustness make it a preferred approach in engineering, economics, and machine learning [10], [11].

2.2.5 *Pseudocode*

```
BEGIN Differential Evolution

    # Define the problem
    Define the objective function f(x)

    # Initialize parameters
    population_size = NP      # Number of individuals in the population
    max_iterations = G        # Maximum number of generations
    F = differential_weight   # Differential weight (usually in [0.5, 1])
    CR = crossover_rate       # Crossover probability (usually in [0, 1])

    # Initialize population
    FOR each individual i in population
        population[i] = InitializeIndividual() # Randomly initialize within the
     problem space
        fitness[i] = f(population[i])          # Evaluate the fitness of each
     individual
    END FOR

    # Main loop
    FOR generation = 1 to max_iterations
        FOR each individual i in population
            # Mutation: Create a mutant vector
            Select three distinct random indices r1, r2, r3 from population such
      that r1 $\neq$ r2 $\neq$ r3 $\neq$ I
            mutant_vector = population[r1] + F * (population[r2] - population[r3
     ])

            # Crossover: Create a trial vector
            trial_vector = []
            FOR each component j in solution vector
                IF Random(0, 1) < CR OR j == random_index THEN
                    trial_vector[j] = mutant_vector[j]
                ELSE
                    trial_vector[j] = population[i][j]
                END IF
            END FOR

            # Selection: Select the better between the trial vector and the
     target vector
            IF f(trial_vector) < fitness[i] THEN
                population[i] = trial_vector
                fitness[i] = f(trial_vector)
            END IF
        END FOR
    END FOR

    # Output the best solution found
    best_individual = population[0]
    best_fitness = fitness[0]
    FOR each individual i in population
        IF fitness[i] < best_fitness THEN
            best_individual = population[i]
            best_fitness = fitness[i]
        END IF
    END FOR

    RETURN best_individual, best_fitness

END Differential Evolution
```

2.2.6 MATLAB® Code

```
function [best_individual, best_fitness] = DifferentialEvolution(f, NP, G, F, CR
    , lower_bound, upper_bound)
    % f: Objective function handle
    % NP: Number of individuals in the population
    % G: Maximum number of generations
    % F: Differential weight
    % CR: Crossover rate
    % lower_bound: Lower bound of the search space
    % upper_bound: Upper bound of the search space

    % Initialize population
    dimensions = length(lower_bound);
    population = lower_bound + (upper_bound - lower_bound) .* rand(NP,
     dimensions);
    fitness = arrayfun(f, population);

    % Main loop
    for generation = 1:G
        for i = 1:NP
            % Mutation: Create a mutant vector
            indices = randperm(NP, 3);
            while any(indices == i)
                indices = randperm(NP, 3);
            end
            r1 = indices(1);
            r2 = indices(2);
            r3 = indices(3);
            mutant_vector = population(r1, :) + F * (population(r2, :) -
     population(r3, :));

            % Ensure the mutant vector is within the bounds
            mutant_vector = max(mutant_vector, lower_bound);
            mutant_vector = min(mutant_vector, upper_bound);

            % Crossover: Create a trial vector
            trial_vector = population(i, :);
            j_rand = randi(dimensions);
            for j = 1:dimensions
                if rand <= CR || j == j_rand
                    trial_vector(j) = mutant_vector(j);
                end
            end

            % Selection: Select the better between the trial vector and the
     target vector
            trial_fitness = f(trial_vector);
            if trial_fitness < fitness(i)
                population(i, :) = trial_vector;
                fitness(i) = trial_fitness;
            end
        end
    end

    % Output the best solution found
    [best_fitness, idx] = min(fitness);
    best_individual = population(idx, :);
end
```

2.3 Particle Swarm Optimization

Particle Swarm Optimization (PSO) [7] is a widely used metaheuristic inspired by flocking behavior. It models search agents that iteratively adjust their positions and velocities using individual experience and collective interactions to find optimal solutions.

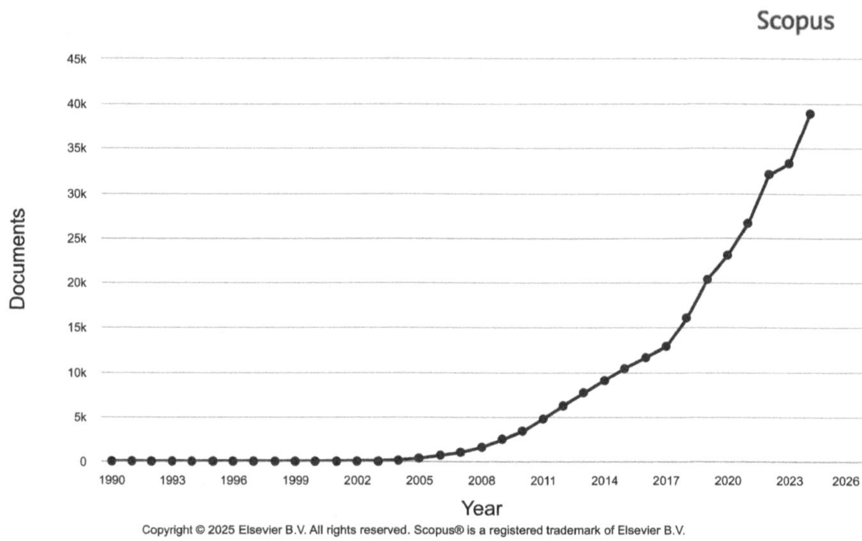

Figure 2.3: Documents by year related to PSO approaches.

2.3.1 *Algorithm Components*

PSO models particle movement through the solution space, adjusting trajectories based on personal and swarm-best positions. The algorithm iterates until reaching an optimal or near-optimal solution. Key steps include:

- Initialization: Generate an initial population of np individuals, each represented as a D-dimensional vector.
- Movement: Update positions based on velocity toward the best-found solutions.
- Inertia Weight balance: Velocity and position updates depend on inertia weight, balancing global exploration and local refinement for optimal search efficiency.

2.3.2 *Particle Position and Velocity*

In Particle Swarm Optimization (PSO), each particle's position in the search space at time t is represented as $x(t)$, while its velocity is $v(t)$. At the next time step, $t+1$, the updated position and velocity $x(t+1)$ and $v(t+1)$, respectively.

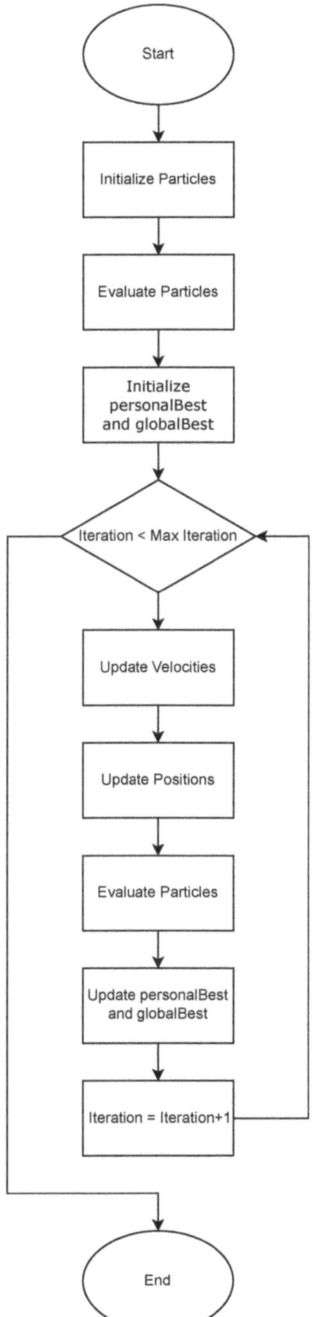

Figure 2.4: Flow diagram of the Particle Swarm Optimization Algorithm.

Two primary factors influence each particle's movement:

Particle's Best-Known Position $p(t)$: The best position the particle has found so far. Global Best Position $g(t)$: The best position discovered by any particle in the swarm.

2.3.3 *Equations Governing Movement*

The velocity and position updates are governed by the following equations:

$$v_i(t+1) = w \cdot v_i(t) + c_1 r_1 (p_i(t) - x_i(t)) + c_2 r_2 (g_i(t) - x_i(t)) \tag{2.4}$$

$$x_i(t+1) = x_i(t) + v_i(t) \tag{2.5}$$

Here, w is the inertia weight, c_1 and c_2 are learning factors that control the influence of $p(t)$ and $g(t)$, respectively, and r_1 and r_2 are random values in the range $[0, 1]$.

2.3.4 *Inertia Weight Strategies*

The inertia weight w plays a crucial role in balancing the global and local search abilities of the particles. Several strategies exist for defining and updating the inertia weight, as shown in Table 2.2.

Table 2.2: Inertia weight strategies.

Strategy	**Definition of Inertia Weight**
Constant inertia weight	$\omega(t) = \omega = const$
Random inertia weight	$\omega(t) = 0.5 + \frac{r}{2} \quad r \sim U(0,1)$
Linearly decreasing inertia weight	$\omega(t) = \omega_{max} - \frac{\omega_{max} - \omega_{min}}{t_{max}} t$
Global-local best inertia weight	$\omega(t) = 1.1 - (\frac{p_{i,j}(t)}{g_j(t)})$
Chaotic descending inertia weight	$\omega(t) = (\omega(0) - \omega(t_{max}))(\frac{t_{max} - t}{t_{max}}) + \omega(t_{max})z$ $z = 4r(1-r)\ with\ r \sim U(0,1)$
Chaotic random inertia weight	$\omega(t) = 0.5r_1 + 0.5z$ $z = 4r_2(1-r_2)\ with\ r_1, r_2 \sim U(0,1)$

2.3.5 *Conclusion and Applications*

Particle Swarm Optimization (PSO) is a versatile and efficient algorithm that leverages the social behavior of particles to solve complex optimization problems. Its simplicity, adaptability, and robust performance make it a popular choice across various fields, including engineering, machine learning, robotics and economics [12], [13].

2.3.6 *Pseudocode*

```
BEGIN Particle Swarm Optimization

    # Define the problem
    Define the objective function f(x)

    # Initialize parameters
    num_particles = N          # Number of particles
    max_iterations = T         # Maximum number of iterations
    w = inertia_weight         # Inertia weight
    c1 = cognitive_constant    # Cognitive (particle) constant
    c2 = social_constant       # Social (swarm) constant

    # Initialize particle positions and velocities
    FOR each particle i in swarm
        position[i] = InitializePosition() # Randomly initialize within the
     problem space
        velocity[i] = InitializeVelocity() # Randomly initialize within a
     defined range
        best_position[i] = position[i]     # Best position for each particle
        best_value[i] = f(position[i])     # Best value for each particle
    END FOR

    # Initialize global best
    global_best_position = best_position[0]
    global_best_value = best_value[0]

    FOR each particle i in swarm
        IF best_value[i] < global_best_value THEN
            global_best_position = best_position[i]
            global_best_value = best_value[i]
        END IF
    END FOR

    # Main loop
    FOR t = 1 to max_iterations
        FOR each particle i in swarm
            # Update velocity
            r1 = Random(0, 1)
            r2 = Random(0, 1)
            velocity[i] = w * velocity[i] +
                          c1 * r1 * (best_position[i] - position[i]) +
                          c2 * r2 * (global_best_position - position[i])

            # Update position
            position[i] = position[i] + velocity[i]

            # Evaluate the new position
            current_value = f(position[i])

            # Update the particle's best known position
            IF current_value < best_value[i] THEN
                best_position[i] = position[i]
                best_value[i] = current_value
            END IF

            # Update the global best known position
            IF current_value < global_best_value THEN
                global_best_position = position[i]
                global_best_value = current_value
            END IF
        END FOR

        # Optionally, adjust parameters such as inertia weight
        w = UpdateInertiaWeight(w, t, max_iterations)

    END FOR
```

```
    # Output the best solution found
    RETURN global_best_position, global_best_value

END Particle Swarm Optimization
```

2.3.7 MATLAB Code

```
function [global_best_individual, global_best_fitness] =
    ParticleSwarmOptimization(f, NP, T, w, c1, c2, lower_bound, upper_bound)
    % f: Objective function handle
    % NP: Number of particles
    % T: Maximum number of iterations
    % w: Inertia weight
    % c1: Cognitive constant
    % c2: Social constant
    % lower_bound: Lower bound of the search space
    % upper_bound: Upper bound of the search space

    % Initialize particle positions and velocities
    dimensions = length(lower_bound);
    position = lower_bound + (upper_bound - lower_bound) .* rand(NP, dimensions)
    ;
    personal_best_position = position;
    velocity = zeros(NP, dimensions);
    fitness = arrayfun(f, position);

    % Initialize global best
    [global_best_fitness, idx] = min(fitness);
    global_best_individual = personal_best_position(idx, :);

    % Main loop
    for t = 1:T
        for i = 1:NP
            % Update velocity
            r1 = rand(1, dimensions);
            r2 = rand(1, dimensions);
            velocity(i, :) = w * velocity(i, :) + ...
                             c1 * r1 .* (personal_best_position(i, :) - position
    (i, :)) + ...
                             c2 * r2 .* (global_best_individual - position(i, :)
    );

            % Update position
            position(i, :) = position(i, :) + velocity(i, :);

            % Ensure the position is within the bounds
            position(i, :) = max(position(i, :), lower_bound);
            position(i, :) = min(position(i, :), upper_bound);

            % Evaluate the new position
            current_fitness = f(position(i, :));

            % Update the particle's best known position
            if current_fitness < fitness(i)
                personal_best_position(i, :) = position(i, :);
                fitness(i) = current_fitness;
            end

            % Update the global best known position
            if current_fitness < global_best_fitness
                global_best_individual = position(i, :);
                global_best_fitness = current_fitness;
            end
```

```
        end

        % Optionally, adjust parameters such as inertia weight
        % w = UpdateInertiaWeight(w, t, T);
    end

    % Output the best solution found
    [global_best_fitness, idx] = min(fitness);
    global_best_individual = position(idx, :);
end
```

References

[1] K. Hussain, M. N. M. Salleh, S. Cheng, and Y. Shi, "Metaheuristic research: a comprehensive survey," *Artificial Intelligence Review*, vol. 52, pp. 2191 – 2233, 2018. [Online]. Available: https://api.semanticscholar.org/CorpusID:32526909

[2] E. Cuevas, D. Zaldívar, and M. Pérez-Cisneros, *Introduction to Metaheuristic Schemes: Characteristics, Properties, and Importance in Solving Optimization Problems*. Cham: Springer Nature Switzerland, 2024, pp. 1–9. [Online]. Available: https://doi.org/10.1007/978-3-031-45561-2_1

[3] K. Sörensen and F. W. Glover, *Metaheuristics*. Boston, MA: Springer US, 2013, pp. 960–970. [Online]. Available: https://doi.org/10.1007/978-1-4419-1153-7_1167

[4] J. Xu and J. Zhang, "Exploration-exploitation tradeoffs in metaheuristics: Survey and analysis," in *Proceedings of the 33rd Chinese Control Conference*, 2014, pp. 8633–8638.

[5] M. Črepinšek, S.-H. Liu, and M. Mernik, "Exploration and exploitation in evolutionary algorithms: A survey," *ACM Comput. Surv.*, vol. 45, no. 3, jul 2013. [Online]. Available: https://doi.org/10.1145/2480741.2480752

[6] R. Storn and K. Price, "Differential evolution – a simple and efficient heuristic for global optimization over continuous spaces," *J. of Global Optimization*, vol. 11, no. 4, p. 341–359, dec 1997. [Online]. Available: https://doi.org/10.1023/A:1008202821328

[7] J. Kennedy and R. Eberhart, "Particle swarm optimization," in *Proceedings of ICNN'95 - International Conference on Neural Networks*, vol. 4, 1995, pp. 1942–1948 vol.4.

[8] S. Das and P. N. Suganthan, "Differential evolution: A survey of the state-of-the-art," *IEEE Transactions on Evolutionary Computation*, vol. 15, pp. 4–31, 2011. [Online]. Available: https://api.semanticscholar.org/CorpusID:206682327

[9] T. M. Shami, A. A. El-Saleh, M. Alswaitti, Q. Al-Tashi, M. A. Summakieh, and S. Mirjalili, "Particle swarm optimization: A comprehensive survey," *IEEE Access*, vol. 10, pp. 10 031–10 061, 2022.

[10] V. Plagianakos, D. Tasoulis, and M. Vrahatis, "A review of major application areas of differential evolution," *Studies in Computational Intelligence*, vol. 143, pp. 197–238, 09 2008.

[11] M. F. Ahmad, N. A. M. Isa, W. H. Lim, and K. M. Ang, "Differential evolution: A recent review based on state-of-the-art works," *Alexandria Engineering Journal*, vol. 61, no. 5, pp. 3831–3872, 2022. [Online]. Available: https://www.sciencedirect.com/science/article/pii/S111001682100613X

[12] J. Fang, W. Liu, L. Chen, S. Lauria, A. Miron, and X. Liu, "A survey of algorithms, applications and trends for particle swarm optimization," *International Journal of Network Dynamics and Intelligence*, vol. 2, no. 1, p. 24–50, Mar. 2023. [Online]. Available: https://www.sciltp.com/journals/ijndi/article/view/176

[13] A. Gad, "Particle swarm optimization algorithm and its applications: A systematic review," *Archives of Computational Methods in Engineering*, vol. 29, p. 2531–2561, 04 2022.

Chapter 3

The Influence of Initialization in Metaheuristics

3.1 Introduction

The initialization stage is crucial in metaheuristic algorithms, setting the foundation for efficient search and exploration within the solution space [1]. It defines initial parameters and generates candidate solutions for subsequent iterations.

Properly configuring algorithm parameters, such as population size, search space bounds, and other algorithm-specific factors, is essential for performance and convergence [2].

Generating initial solutions is key to the optimization process and varies by problem. A common approach is random initialization within search space boundaries, promoting exploration and minimizing suboptimal solutions, enhancing overall robustness.

Beyond random generation, heuristic techniques like problem-specific rule-based methods leverage prior knowledge or heuristic insights to create high-quality initial solutions [3]. Effective initialization accelerates optimization, while poor initialization risks premature convergence or suboptimal search outcomes.

The initialization stage in metaheuristic algorithms is essential for defining parameters and generating candidate solutions. Ongoing research explores new techniques to enhance initialization, improving the discovery of optimal or near-optimal solutions in complex optimization tasks.

DOI: 10.1201/9781032723563-3

This chapter examines the impact of initialization in metaheuristic algorithms and explores various techniques for generating initial candidate solutions. It presents alternatives to common pseudo-random methods and provides an in-depth discussion of several algorithms, offering valuable insights for researchers, students, and enthusiasts in the field.

3.2 Key Findings of the Analyzed Papers

The growth graph from the Scopus database illustrates a rising trend in published papers on initialization in metaheuristic algorithms. Between 2002 and 2016, annual publications remained low (under 50), but since 2016, and especially post-2020, exponential growth has occurred, surpassing 350 papers in 2024. This surge reflects increasing interest, likely driven by technological advancements and applications in Machine learning [4], Deep learning [5] or Big data [6].

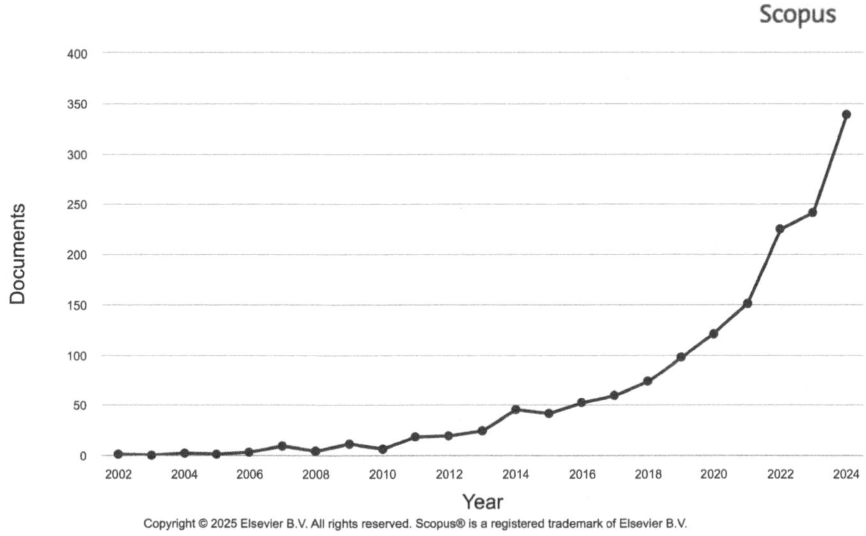

Figure 3.1: Documents by year related to initialization approaches.

Effective initialization in metaheuristics enhances search quality and efficiency in complex optimization problems. As modern applications demand tailored algorithms, refining initialization strategies remains essential. The following sections examine various initialization methods, analyzing their distinct approaches.

3.2.1 *Chaos-based Initialization*

Chaos theory plays a key role in metaheuristic initialization, leveraging sensitivity to initial conditions to generate diverse solutions. Chaotic maps enhance exploration with deterministic yet unpredictable patterns, benefiting algorithms like PSO and DE, particularly in constrained optimization. While effective in balancing exploration and exploitation, improper parameter tuning can lead to suboptimal performance [3].

3.2.2 *Clustering-based Methods*

Clustering algorithms enhance diversity by partitioning the search space into regions, improving initial coverage, especially in multi-objective optimization. While effective, their computational overhead can be limiting in time-sensitive applications. [3] highlights clustering's benefits in search space coverage but notes its higher computational cost.

3.2.3 *Low-Discrepancy Sequences (LDS)*

Quasi-random sequences like Halton, Sobol, and Latin Hypercube enhance initial population uniformity, improving solution accuracy without harming algorithm performance. Sobol sequences show superior convergence in GAs and PSO, though LDS-based initialization can vary based on problem dimensionality and structure [7, 8].

3.2.4 *Heuristic-based Initialization*

Heuristic initialization utilizes problem-specific knowledge to generate strong starting solutions, improving PSO performance in task scheduling. While effective in specific domains, excessive reliance may limit generalizability and risk premature convergence, as explored in [9].

3.2.5 *Hybrid and Novel Approaches*

Hybrid methods integrate multiple strategies to address individual technique limitations. DLU initialization partitions the search space, boosting convergence and accuracy in DE and PSO. The Confluence Metaheuristic mimics river confluence dynamics, enhancing diversity and solution quality in continuous optimization. These approaches, validated in benchmark studies [10, 11], show promise for overcoming traditional limitations.

3.3 Comparative Analysis

To illustrate the effectiveness of these initialization methods, Table 3.1 summarizes their performance based on recent studies.

Table 3.1: Performance of initialization methods based on recent studies.

Initialization Method	Strengths	Weaknesses	Key Applications
Chaos-Based	High diversity, sensitive to initial conditions	Requires careful parameter tuning	Constrained and discrete problems
Clustering	Enhanced search space coverage	High computational cost	Population-based algorithms
Low-discrepancy sequences	Uniform distribution, robust performance	May not always outperform random	Benchmark optimization problems
Heuristic-based	Problem-specific efficiency	Risk of reduced exploration	Task scheduling, TSP
Hybrid	Combines diversity and efficiency	Implementation complexity	Complex optimization problems

3.4 Statistical Insights

This chapter examines studies on the impact of initialization in metaheuristic optimization algorithms. Selected works were analyzed based on their research focus and methodology. A comparative table summarizes key indicators, providing a clearer view of each approach's unique characteristics.

Table 3.2 compares performance across benchmark functions, emphasizing.

The table offers a comparative analysis of algorithms and initialization methods, highlighting their strengths and limitations. Heuristic and hybrid approaches typically ensure faster convergence and high solution quality but may need problem-specific tuning.

3.5 Discussion

Effective initialization balances exploration (diversity) and exploitation (intensification). Algorithms like PSO, ABC, and SFS are highly sensitive to initialization, whereas DE is more robust. Advanced methods must offer substantial performance gains to justify their computational costs.

Future research should prioritize hybrid approaches that adapt initialization strategies to problem complexity, leverage machine learning for predicting optimal methods, and explore multi-population and cooperative metaheuristics for improved optimization.

The initialization method strongly affects metaheuristic algorithm performance. Leveraging diverse, problem-specific strategies enhances efficiency and robustness, especially for complex optimization tasks.

Table 3.2: Comparative performance evaluation of initialization methods.

Algorithm	**Initialization Method**	**Convergence Speed**	**Solution Quality**	**Initialization Sensitivity**	**Key Observations**
PSO	Chaos-Based	High	Good	High	Effective for constrained optimization but requires careful parameter tuning [12].
PSO	Heuristic (LJFP, MCT)	Very High	Excellent	Moderate	Optimized for task scheduling problems; balances exploration and exploitation [9].
DE	Diagonal Linear Uniform (DLU)	Very High	Excellent	Low	Consistently improves convergence speed and solution accuracy across benchmarks [10].
SFS	LDS	Moderate	High	High	Sobol and Latin Hypercube perform well for multi-modal and complex problems [7].
Genetic Algorithm	Heuristic	Very High	Excellent	Moderate	The proposed technique depends on the linear regression technique [13].
ABC	Clustering-Based	Moderate	Good	High	Enhances diversity at the cost of increased computational requirements [3].
Confluence	Confluence Metaheuristic	High	Excellent	Low	Novel approach that improves solution diversity by simulating river confluence dynamics [11].
Simulated Annealing	Random	Low	Moderate	Moderate	Works well for small-scale problems but struggles with scalability [14].
Harmony Search	Random	Very High	Excellent	Low	Effective for multi-modal problems with robust global search capabilities [15].

References

[1] M. Abdel-Basset, L. Abdel-Fatah, and A. K. Sangaiah, "Metaheuristic algorithms: A comprehensive review," *Computational Intelligence for Multimedia Big Data on the Cloud with Engineering Applications*, pp. 185–231, 2018.

[2] A. H. Gandomi, X.-S. Yang, S. Talatahari, and A. H. Alavi, "Metaheuristic algorithms in modeling and optimization," *Metaheuristic Applications in Structures and Infrastructures*, vol. 1, pp. 1–24, 2013.

[3] M. Sarhani, S. Voß, and R. Jovanovic, "Initialization of metaheuristics: comprehensive review, critical analysis, and research directions," *International Transactions in Operational Research*, vol. 30, no. 6, pp. 3361–3397, 2023.

[4] B. Mahesh, "Machine learning algorithms-a review," *International Journal of Science and Research (IJSR).[Internet]*, vol. 9, no. 1, pp. 381–386, 2020.

[5] Y. LeCun, Y. Bengio, and G. Hinton, "Deep learning," *nature*, vol. 521, no. 7553, pp. 436–444, 2015.

[6] A. Oussous, F.-Z. Benjelloun, A. A. Lahcen, and S. Belfkih, "Big data technologies: A survey," *Journal of King Saud University-Computer and Information Sciences*, vol. 30, no. 4, pp. 431–448, 2018.

[7] M. A. Navarro, D. Oliva, A. Ramos-Michel, B. Morales-Castaneda, D. Zaldívar, and A. Luque-Chang, "A review of the use of quasi-random number generators to initialize the population in meta-heuristic algorithms," *Archives of Computational Methods in Engineering*, vol. 29, no. 7, pp. 5149–5184, 2022.

[8] J. O. Agushaka and A. E. Ezugwu, "Initialization approaches for population-based metaheuristic algorithms: a comprehensive review," *Applied Sciences*, vol. 12, no. 2, p. 896, 2022.

[9] S. A. Alsaidy, A. D. Abbood, and M. A. Sahib, "Heuristic initialization of pso task scheduling algorithm in cloud computing," *Journal of King Saud*

University-Computer and Information Sciences, vol. 34, no. 6, pp. 2370–2382, 2022.

[10] Q. Li, Y. Bai, and W. Gao, "Improved initialization method for metaheuristic algorithms: a novel search space view," *IEEE Access*, vol. 9, pp. 121 366–121 384, 2021.

[11] D. Kaushik and M. Nadeem, "Confluence metaheuristic: a novel initialization strategy for metaheuristic algorithms," *Evolving Systems*, vol. 15, no. 2, pp. 429–454, 2024.

[12] D. Tian, "Particle swarm optimization with chaos-based initialization for numerical optimization," *Intelligent Automation & Soft Computing*, pp. 1–12, 2017.

[13] A. B. Hassanat, V. S. Prasath, M. A. Abbadi, S. A. Abu-Qdari, and H. Faris, "An improved genetic algorithm with a new initialization mechanism based on regression techniques," *Information*, vol. 9, no. 7, p. 167, 2018.

[14] E. Celik and D. Dal, "A novel simulated annealing-based optimization approach for cluster-based task scheduling," *Cluster Computing*, vol. 24, no. 4, pp. 2927–2956, 2021.

[15] Z. W. Geem, J. H. Kim, and G. V. Loganathan, "A new heuristic optimization algorithm: harmony search," *simulation*, vol. 76, no. 2, pp. 60–68, 2001.

Chapter 4

Different Methodologies for Initialization

4.1 Introduction

Global optimization algorithms have been widely studied due to their broad applications. Stochastic algorithms, central to applied mathematics and numerical analysis, efficiently find global optima in complex hyperspaces. Despite an evolving taxonomy, they are often categorized by their underlying concepts, fields, or inspiration.

Various optimization methods draw inspiration from evolution, such as Evolutionary Strategies (ES) [1], Genetic Algorithms (GA) [2], and Differential Evolution (DE) [3]. Physical phenomena have led to algorithms like Gravitational Search Algorithm (GSA) [4], Prism Refraction Search (PRS) [5], and Stochastic Fractal Search (SFS) [6]. Quasi-Random Fractal Search (QRFS) [7], Harmony Search (HS) [8], the Equilibrium Optimizer (EO) [9]. Mathematical principles underpin techniques like Covariance Matrix Adaptation Evolution Strategy (CMA-ES) [10], the Sine Cosine Algorithm (SCA) [11], and Archimedes Optimization Algorithm (AOA) [12]. Social behaviors inspire methods such as Teaching-Learning-Based Optimization (TLBO) [13], Hyper Particle Swarm Optimization (HyperPSO) [14], Cultural Algorithms (CA) [15], Particle Swarm Optimization (PSO) [16], and Ant Colony Optimization (ACO) [17]). Other notable approaches include Whale Optimization Algorithm (WOA) [18], Crow Search Algorithm (CSA) [19], Side-Blotched Lizard Algorithm (SBLA) [20], Harris Hawks Optimization (HHO) [21], and the Aquila Optimizer (AO) [22], among others.

In global optimization algorithms, the process begins by randomly sampling N individuals of dimension D, each representing a candidate solution. Studies highlight the impact of spatial distribution on algorithm performance [23, 24, 25, 26, 27, 28,

DOI: 10.1201/9781032723563-4

29]. Various initialization techniques exist, categorized by their effectiveness and justification for use.

4.2 Initialization Techniques

4.2.1 *Random Initialization*

Random initialization is a widely used metaheuristic strategy, generating uniformly distributed solutions within the search space [30].

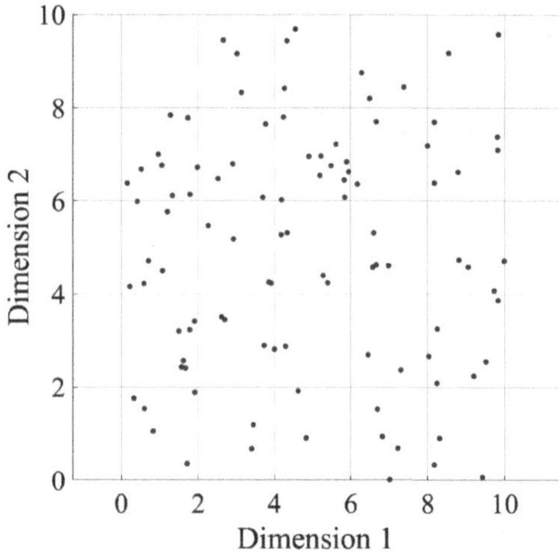

Figure 4.1: Random initialization in 2D.

This method enhances diversity in initial solutions, preventing premature bias in the search space [31]. However, it may suffer from inefficient convergence, especially in large or complex spaces where random solutions are distant from optimal regions.

T following provides a concise explanation of random initialization, along with an illustrative example in pseudocode for better understanding.

First, the initialization process is divided into data inputs and outputs for clarity and efficiency.

Inputs:

n: Number of solutions to generate (population size). **d**: Dimensionality of the problem (number of variables per solution). $Xmin$ and $Xmax$: Lower and upper bounds of the search space for each dimension.

Outputs:

Pop: Initial set of random solutions.

4.2.1.1 Pseudocode 1

```
BEGIN Random Initialization

    # Define the problem
    Define the number of variables d
    Define the lower bounds Xmin and upper bounds Xmax for each variable

    # Initialize parameters
    population_size = n     # Number of individuals (solutions) in the population

    # Initialize population
    FOR each individual i in population
        FOR each variable j in solution vector
            Pop[i][j] = Random(Xmin[j], Xmax[j])  # Generate a random value for
     each variable
        END FOR
    END FOR

    # Output the initialized population
    RETURN Pop

END Random Initialization
```

Pseudocode description

- Define the problem: Specify the number of variables (d) and set the lower (Xmin) and upper (Xmax) bounds for each within the search space.
- Initialize parameters: Set the population size (n), determining the number of randomly generated solutions.
- Population initialization: Each solution is assigned random values within Xmin[j] and Xmax[j], ensuring it stays within the defined search space.
- Output: The generated initial population is returned, ready for use in the metaheuristic algorithm.

4.2.1.2 Illustrative Example of Random Initialization

To initialize a three-solution population for a two-variable optimization problem, set the search space bounds:

- Xmin = [0, 0] (both variables have a lower bound of 0).
- Xmax = [10, 20] (first variable upper bound 10, second variable upper bound 20).

Algorithm application:

- n = 3 (number of solutions)
- d = 2 (number of variables)

The algorithm generates three random solutions within the given ranges:

Solution 1: [5.32, 15.87]

Solution 2: [7.90, 3.76]

Solution 3: [2.48, 12.67]

These solutions form the initial *Pop* set for the metaheuristic algorithm.

4.2.1.3 Advantages and Disadvantages of Random Initialization

Table 4.1 compares the advantages and disadvantages of random initialization in optimization algorithms.

Table 4.1: Overview of the pros and cons of random initialization in optimization algorithms.

Advantages	Disadvantages
Simplicity: Easy to implement, requiring no extra problem-specific information.	*Slow convergence*: Random solutions may start far from the optimum, delaying the convergence process.
Diversity: Random solutions ensure a broad distribution within the search space, fostering effective exploration.	*Lack of initial exploitation*: This approach does not utilize problem-specific information, making it less efficient than heuristic or knowledge-based methods that start with solutions closer to the optimum.

4.2.2 MATLAB® Code Example of Random Initialization

```
function Pop = random_initialization(population_size, dim, Xmin, Xmax)
    % RANDOM_INITIALIZATION Generates a randomly initialized population.
    % Inputs:
    %   population_size - Number of individuals (solutions) in the population
    %   dim - Number of variables (dimensions)
    %   Xmin - Lower bounds for each variable (vector of size dim)
    %   Xmax - Upper bounds for each variable (vector of size dim)
    % Output:
    %   Pop - Matrix of size (population_size x dim) with initialized values

    % Initialize population matrix
    Pop = zeros(population_size, dim);

    % Generate random values for each variable within the given bounds
    for i = 1:population_size
        for j = 1:dim
            Pop(i, j) = Xmin(j) + rand() * (Xmax(j) - Xmin(j));
        end
    end
end
```

4.2.3 *Heuristic Initialization*

Heuristic Initialization generates the initial population using domain-specific heuristics rather than random methods [32], [33]. This approach produces higher-quality solutions closer to the optimal region, enhancing convergence speed. By leveraging problem-specific knowledge, it directs the search toward promising areas, minimizing exposure to poor-quality regions.

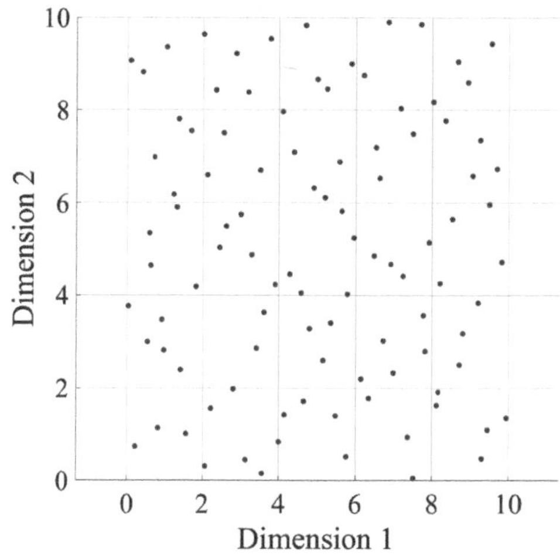

Figure 4.2: Heuristic initialization using Halton sequence in 2D.

This method is valuable when prior knowledge or patterns help generate solutions, enabling faster exploration and efficient exploitation of the search space.

General approach

The choice of heuristic for initialization depends on the problem. For combinatorial optimization (e.g., Traveling Salesman Problem (TSP) [34]), greedy heuristics generate near-optimal solutions. In continuous optimization, heuristics leverage derivatives, rules of thumb, or constraints to guide initialization.

The initialization process is divided into data inputs and outputs:

Inputs:

n: Number of solutions (population size).
d: Problem dimensionality (decision variables).
HeuristicMethod(): Domain-specific function for solution generation.

Outputs:

Pop: Set of heuristic-generated initial solutions.

4.2.3.1 Pseudocode 1

```
BEGIN Heuristic Initialization

  # Define the problem
  Define the number of variables d
  Define the heuristic method HeuristicMethod()

  # Initialize parameters
  population_size = n     # Number of individuals (solutions) in the population

  # Initialize population
  FOR each individual i in population
      solution_i = HeuristicMethod()   # Generate solution using the heuristic
      Add solution_i to Pop            # Add the generated solution to the
  set
  END FOR

  # Return the heuristic-initialized population
  RETURN Pop

END Heuristic Initialization
```

Pseudocode explanation

- Input parameters: The function requires n (number of solutions), d (decision variables), and a heuristic function (HeuristicMethod()) to generate domain-specific solutions. *Pop* is initialized to store the results.
- Generating solutions: A loop runs n times to generate each solution. For each iteration:
  - The heuristic method (HeuristicMethod()) is called to generate a solution.
  - The generated solution is added to the set *Pop*.
- Returning the population: Once all n solutions have been generated using the heuristic, the population *Pop* is returned.

4.2.3.2 Illustrative Example of Heuristic Initialization in TSP

In the Traveling Salesman Problem (TSP), a common heuristic is the nearest neighbor heuristic. This heuristic starts at a random city and then iteratively selects the nearest unvisited city until all cities have been visited. Here's how this would look in Pseudocode of Subsubsection: 4.2.3.3:

4.2.3.3 Pseudocode 2

```
BEGIN Nearest Neighbor TSP Heuristic

  # Initialize the tour
  unvisited = set of all cities except start_city
  tour = [start_city]
  current_city = start_city

  # Construct the tour
  WHILE unvisited is not empty
      next_city = city in unvisited closest to current_city
      tour.append(next_city)
```

```
        unvisited.remove(next_city)
        current_city = next_city
    END WHILE

    # Return the constructed tour
    RETURN tour

END Nearest Neighbor TSP Heuristic
```

4.2.3.4 Pseudocode 3

```
BEGIN Heuristic Initialization for TSP

    # Define the problem
    Define the set of cities

    # Initialize population
    FOR each individual i in population
        start_city = Select a random city from cities
        solution_i = Nearest_Neighbor_TSP_Heuristic(cities, start_city)
        Add solution_i to Pop  # Add the generated solution to the population
    END FOR

    # Output the initialized population
    RETURN Pop

END Heuristic Initialization for TSP
```

These pseudocodes outline the process of generating an initial population using heuristic-based methods, including a specific example for the Traveling Salesman Problem (TSP) using the nearest neighbor heuristic.

4.2.3.5 Advantages and Disadvantages of Heuristic Initialization

Heuristic initialization provides a structured alternative to random initialization, using domain-specific knowledge to guide the search effectively. However, it has its own pros and cons, summarized in Table 4.2.

Table 4.2: Comparison of pros and cons of heuristic initialization methods.

Pros	Cons
Faster convergence: Starting with high-quality solutions accelerates progress toward optimal or near-optimal results.	*Limited diversity*: Heuristics follow predefined strategies, which may reduce solution variety, risking premature convergence and poor exploration.
Problem-specific knowledge: Heuristics use domain expertise to steer the search toward promising regions, improving performance.	*Implementation complexity*: Heuristic design demands domain knowledge and is more complex than random initialization.
Avoiding of poor regions: Heuristic initialization steers the search away from low-quality areas, improving efficiency over random methods.	*Search bias*: Heuristic methods may focus too heavily on certain areas, risking the omission of other promising regions.

4.2.4 MATLAB Code Examples of Heuristic Initialization

```
function Pop = heuristic_initialization(population_size, d, HeuristicMethod)
    % HEURISTIC_INITIALIZATION - Initializes a population using a heuristic
    method
    % Inputs:
    %   population_size - Number of individuals (solutions) in the population
    %   d - Number of variables (dimensions) in the solution
    %   HeuristicMethod - Function handle for the heuristic method
    % Output:
    %   Pop - Initialized population matrix

    % Initialize population matrix
    Pop = zeros(population_size, d);

    % Generate solutions using the heuristic method
    for i = 1:population_size
        Pop(i, :) = HeuristicMethod(d);
    end
end
```

```
function tour = nearest_neighbor_tsp(distanceMatrix, start_city)
    % Nearest Neighbor TSP Heuristic
    % distanceMatrix: distance matrix between cities
    % start_city: starting city (index)
    % tour: sequence of cities in the tour

    numCities = size(distanceMatrix, 1); % Total number of cities
    unvisited = 1:numCities; % List of unvisited cities
    unvisited(unvisited == start_city) = []; % Remove the starting city

    tour = start_city; % Initialize the tour with the starting city
    current_city = start_city;

    % Construct the tour
    while ~isempty(unvisited)
        % Find the nearest unvisited city
        [~, idx] = min(distanceMatrix(current_city, unvisited));
        next_city = unvisited(idx);

        % Add the city to the tour and update lists
        tour = [tour, next_city];
        unvisited(unvisited == next_city) = [];
        current_city = next_city;
    end

    % Close the circuit by returning to the starting city
    tour = [tour, start_city];
end
```

```
% Heuristic Initialization for TSP in MATLAB

function Pop = heuristicInitializationTSP(numIndividuals, cities, distanceMatrix
    )
    % Initialize population
    Pop = cell(numIndividuals, 1);

    for i = 1:numIndividuals
        startCity = randi(length(cities)); % Select a random city
        solution = nearestNeighborTSP(cities, startCity, distanceMatrix);
        Pop{i} = solution; % Add the generated solution to the population
    end
end

function tour = nearestNeighborTSP(cities, startCity, distanceMatrix)
    % Nearest Neighbor TSP Heuristic
    unvisited = 1:length(cities);
```

```
    unvisited(startCity) = []; % Remove start city from unvisited list
    tour = [startCity]; % Initialize tour with start city
    currentCity = startCity;

    while ~isempty(unvisited)
        % Find the closest city
        [~, idx] = min(distanceMatrix(currentCity, unvisited));
        nextCity = unvisited(idx);
        tour = [tour, nextCity]; % Append to tour
        unvisited(idx) = []; % Remove from unvisited
        currentCity = nextCity;
    end
end
```

4.2.5 *Knowledge-based Initialization*

Knowledge-based initialization utilizes prior knowledge (e.g., experience, empirical data, expert insights) to generate initial solutions, enhancing solution quality and convergence speed in metaheuristic algorithms [35].

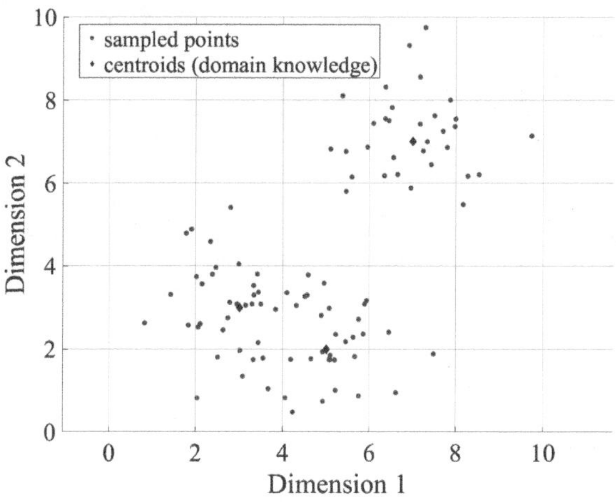

Figure 4.3: Knowledge-based initialization in 2D.

This method is effective when solution space aspects are well-understood or historical data is available, guiding the algorithm to promising regions early, reducing time and computational effort for optimal solutions.

Characteristics

- Data-driven: Uses empirical data or past solutions to generate initial candidates efficiently.
- Expert-driven: Uses domain expertise to generate solutions aligned with optimal or feasible search space regions.
- Hybrid approaches: Integrate heuristics to refine solutions using problem constraints or patterns for better optimization.

The initialization procedure is structured into data inputs and outputs, detailing key steps for implementation.

Inputs:

n: Number of generated solutions (population size).
d: Problem dimensionality or number of decision variables.
KnowledgeBase(): A function that retrieves or computes initial solutions based on domain knowledge or historical data.

Outputs:

Pop: Set of initial solutions generated using knowledge-based techniques.

4.2.5.1 Pseudocode 1

```
BEGIN Knowledge-Based Initialization

  # Define the problem
  Define the number of variables d
  Define the knowledge base method KnowledgeBase()

  # Initialize parameters
  population_size = n    # Number of individuals (solutions) in the population

  # Initialize population
  FOR each individual i in population
      solution_i = KnowledgeBase()  # Retrieve or compute a solution using
   prior knowledge
      Add solution_i to Pop            # Add the generated solution to the set
  END FOR

  # Output the initialized population
  RETURN Pop

END Knowledge-Based Initialization
```

Pseudocode explanation

- Input parameters: The function takes n (solutions), d (decision variables), and KnowledgeBase(), which retrieves or generates solutions using prior knowledge or historical data. P is initialized to store the results.
- Solution generation: A loop runs n times to generate each solution. For each iteration:
  - KnowledgeBase() generates or retrieves solutions using past optimal data, known rules, patterns, or expert-driven techniques for improved initialization.
  - The generated solution is then added to the set Pop.
- Returning the population: After generating n solutions with knowledge-based methods, the population P is returned.

4.2.5.2 Illustrative Example for Knowledge-based Initialization in Production Scheduling

Historical data-driven scheduling: Past machine performance, job processing times, and bottlenecks help generate near-optimal initial schedules in production planning.

4.2.5.3 Pseudocode 2: Example of Production Scheduling Knowledge-based Initialization

```
BEGIN Retrieve Historical Schedule

    # Example knowledge: retrieve past schedules or rules
    schedule = Load past schedule from data
    Adjust schedule based on current production needs and constraints

    RETURN schedule

END Retrieve historical schedule


BEGIN Knowledge-Based Initialization for Scheduling

    # Define the problem
    Define the set of jobs and machines

    # Initialize population
    FOR each individual i in population
        solution_i = Retrieve_Historical_Schedule()  # Use knowledge from past
     scheduling data
        Add solution_i to Pop                         # Add the generated
     schedule to the population
    END FOR

    # Output the initialized population
    RETURN Pop

END Knowledge-Based Initialization for Scheduling
```

Retrieve_Historical_Schedule() pulls past scheduling data, adjusts for current constraints, and incorporates proven practices, creating an initial population of feasible, near-optimal schedules, enhancing metaheuristic performance.

Variations of Knowledge-Based initialization include:

Empirical data-driven [36]: Uses past solution data for guidance, e.g., historical routing in vehicle routing problems.

Expert-driven rules [37]: Generates solutions using domain expertise, e.g., predefined operation sequences in workshop scheduling.

Pre-optimized solutions [38]: Reuses simpler or past runs to initialize solutions, e.g., leveraging previous scenarios in mobile robot route planning.

Constraint-based knowledge [39], [40]: Incorporates known constraints (legal, environmental, physical), e.g., ensuring safety and stability in structural engineering optimization.

4.2.5.4 Pros and Cons of Knowledge-based Initialization

Knowledge-based initialization methods leverage historical data, expert insights, and domain knowledge to generate high-quality initial solutions for optimization algorithms. While it enhances convergence speed and solution quality, it also has limitations, summarized in Table 4.3.

Table 4.3: Comparison of pros and cons of knowledge-based initialization methods.

Pros	Cons
High-quality initial solutions: Knowledge-based methods yield superior solutions compared to random or heuristic initialization.	*Limited exploration*: Knowledge-based initialization favors known regions, potentially restricting exploration, causing premature convergence, or missing better solutions.
Faster convergence: Starting with high-quality solutions accelerates progress toward optimal or near-optimal results, reducing iterations.	*Implementation complexity*: Requires domain expertise and may involve data collection and processing, adding to design challenges.
Leveraging past experience: Algorithms learn from historical data or expert knowledge, benefiting industries with well-defined processes.	*Dependence on data quality*: Outdated or low-quality prior knowledge can bias the algorithm toward suboptimal search regions.

4.2.6 MATLAB Code Examples of Knowledge-based Initialization

```
function Pop = knowledgeBasedInitialization(n, d)
    % Knowledge-Based Initialization in MATLAB
    % n: Population size
    % d: Number of variables (dimensions)

    Pop = zeros(n, d); % Initialize population matrix

    for i = 1:n
        solution_i = knowledgeBase(d); % Retrieve or compute solution
        Pop(i, :) = solution_i; % Add solution to population
    end
end

function solution = knowledgeBase(d)
    % Example knowledge-based method
    % Modify this function to use actual domain knowledge

    solution = rand(1, d); % Placeholder: random solution in [0,1]^d
end
```

```
function Pop = knowledgeBasedInitializationForScheduling(population_size)
    % Knowledge-Based Initialization for Scheduling
    % Define the problem: set of jobs and machines

    Pop = cell(population_size, 1); % Initialize population as a cell array

    for i = 1:population_size
        solution_i = retrieveHistoricalSchedule(); % Use past scheduling data
        Pop{i} = solution_i; % Add the generated schedule to the population
    end

end

function schedule = retrieveHistoricalSchedule()
```

```
    % Retrieve Historical Schedule
    % Example: Load past schedule from data

    % Simulating loading past schedule (replace with actual data retrieval)
    schedule = loadPastSchedule();

    % Adjust schedule based on current production needs and constraints
    schedule = adjustSchedule(schedule);

end

function schedule = loadPastSchedule()
    % Dummy function to simulate loading a past schedule
    schedule = randperm(10); % Example: random permutation of 10 jobs
end

function adjustedSchedule = adjustSchedule(schedule)
    % Dummy function to adjust schedule based on constraints
    adjustedSchedule = schedule; % Here, you could apply adjustments
end
```

4.2.7 *Deterministic Initialization*

Deterministic initialization [41] generates predictable, non-random solutions, ensuring consistency and repeatability. Unlike random or heuristic methods, it always produces identical populations for the same inputs, aiding full search space coverage via structured distribution.

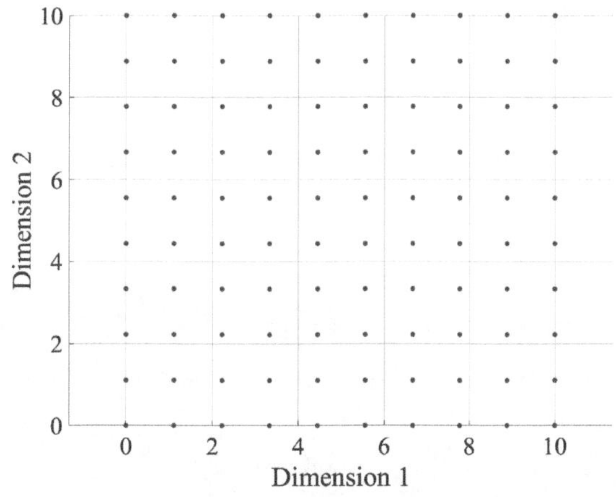

Figure 4.4: Deterministic initialization in 2D.

Deterministic initialization is useful in structured problem spaces or where key search regions require thorough exploration, ensuring regularity and uniformity in the initial population.

Characteristics

- Consistency and repeatability: With identical parameters, the same initial population is always produced, ensuring predictability.

- Structured exploration: Solutions are systematically distributed across the search space for comprehensive coverage.
- Search space coverage: Deterministic initialization ensures even distribution, preventing clustering or gaps in key regions.

Popular deterministic initialization methods:

Grid-based initialization: Divides the search space into a grid, placing solutions at regular intervals for even distribution.

Low-discrepancy sequences:

Latin Hypercube Sampling (LHS) [42]: Divides the search space into subregions, selecting points to ensure uniform coverage.

Sobol sequences [43]: Minimizes clustering, distributing points uniformly for improved search space exploration.

As with previous initialization types, the method is outlined in terms of inputs and outputs for clarity in operation.

Inputs:

n: Number of generated solutions (population size).
d: Problem dimensionality or number of decision variables.
InitializationMethod(): Generates solutions using a deterministic rule or pattern (e.g., grid, Sobol sequence).

Outputs:

Pop: Initial solutions generated via deterministic techniques.

4.2.7.1 Pseudocode 1

```
BEGIN Deterministic Initialization

    # Define the problem
    Define the number of variables d
    Define the method InitializationMethod() for generating deterministic
     solutions

    # Initialize parameters
    population_size = n  # Number of individuals (solutions) in the population

    # Initialize population using deterministic rules
    FOR each individual i in population
        solution_i = InitializationMethod(i)  # Generate solution using a
     deterministic rule or pattern
        Add solution_i to Pop                 # Add the generated solution to
     the set
    END FOR

    # Output the initialized population
    RETURN Pop

END Deterministic Initialization
```

Pseudocode explanation

- Input parameters: The function specifies n (solutions), d (decision variables), and a deterministic method (InitializationMethod()) that follows a predefined rule or pattern. *Pop* is initialized to store the generated solutions.
- Generating solutions: A loop iterates n times, calling InitializationMethod() to generate solutions deterministically, ensuring even spacing or adherence to a structured pattern.
- Returning the population: After generating n solutions via a deterministic method, the population Pop is returned.

4.2.7.2 Example for Grid-based Initialization

Grid-based initialization: Divides the search space into equal intervals, placing solutions at grid points for even distribution.

4.2.7.3 Pseudocode 2

```
BEGIN Grid-Based Initialization

    # Define the problem
    Define the number of variables d
    Define the lower bounds Xmin and upper bounds Xmax for each variable

    # Initialize parameters
    population_size = n        # Number of individuals in the population
    grid_points_per_dimension = n^(1/d)  # Calculate how many points per
     dimension are needed to cover the space

    # Initialize population
    FOR each grid point in the d-dimensional space
        solution_i = []
        FOR each variable j in solution vector
            solution_i[j] = Xmin[j] + (Xmax[j] - Xmin[j]) * Grid_Position(j) / (
     grid_points_per_dimension - 1)
        END FOR
        Add solution_i to Pop  # Add the generated solution to the population
    END FOR

    # Output the initialized population
    RETURN Pop

END Grid-Based Initialization
```

Grid-based approach: Divides the search space into equal intervals along each dimension, placing solutions at grid points based on population size and problem dimensionality.

Pros of Grid-based initialization

- Ensures uniform coverage of the search space.
- Provides a structured, systematic approach to search space exploration.

Cons of Grid-based initialization

- In high-dimensional problems, the grid size expands exponentially, reducing efficiency.

4.2.7.4 Example Pseudocode for Sobol Sequence Initialization

4.2.7.5 Pseudocode 3

```
BEGIN Sobol Sequence Initialization

    # Define the problem
    Define the number of variables d
    Define the lower bounds Xmin and upper bounds Xmax for each variable

    # Initialize parameters
    population_size = n  # Number of individuals in the population
    sobol_sequence_generator = Initialize_Sobol_Sequence_Generator(d)

    # Initialize population
    FOR each individual i in population
        sobol_point = sobol_sequence_generator()  # Generate a Sobol sequence
     point
        solution_i = []
        FOR each variable j in solution vector
            solution_i[j] = Xmin[j] + (Xmax[j] - Xmin[j]) * sobol_point[j]
        END FOR
        Add solution_i to Pop  # Add the generated solution to the population
    END FOR

    # Output the initialized population
    RETURN Pop

END Sobol Sequence Initialization
```

Sobol sequences: Generate points that uniformly fill the search space, scaling each to fit within defined problem bounds for decision variables.

Pros of Sobol sequences

- Provide uniform coverage of the search space.
- Work well in higher dimensions, offering better exploration compared to grid-based methods.

Cons of Sobol sequences

- Require additional computational effort to generate Sobol points compared to simpler methods like random initialization.

4.2.7.6 Pros and Cons of Deterministic Initialization

Deterministic initialization methods systematically generate initial solutions, ensuring even search space coverage and repeatability, beneficial in structured problem domains. However, they face challenges like limited diversity and inefficiency in high-dimensional problems. See Table 4.4 for pros and cons.

Table 4.4: Comparison of pros and cons of deterministic initialization methods.

Pros	Cons
Full coverage of the search space: Deterministic initialization systematically generates solutions, ensuring even and full search space coverage, benefiting well-structured problem spaces.	*Limited diversity risk*: Poorly designed deterministic methods may miss solutions in non-grid areas or favor specific regions, reducing search effectiveness.
Repeatability: The deterministic method ensures the same population for identical inputs, enabling consistent and comparable experiments.	*High dimensionality challenges*: Grid-based approaches face exponential growth in points as the number of variables increases, reducing efficiency.
Bias elimination: Deterministic methods minimize clustering and reduce bias risks inherent in random initialization.	*Less flexibility*: Deterministic methods struggle in dynamic or stochastic environments, where randomness aids exploration.

4.2.8 *MATLAB Code Examples of Deterministic Initialization*

```
function Pop = deterministicInitialization(n, d)
    % Deterministic Initialization for an Optimization Problem
    %
    % Inputs:
    %   n - Population size (number of individuals)
    %   d - Number of variables (dimensions)
    %
    % Output:
    %   Pop - Initialized population using a deterministic rule

    % Initialize empty population matrix
    Pop = zeros(n, d);

    % Generate deterministic solutions
    for i = 1:n
        Pop(i, :) = initializationMethod(i, d);
    end
end

function solution = initializationMethod(index, d)
    % Example deterministic initialization method
    % Generates a sequence based on a linear distribution
    %
    % Inputs:
    %   index - Individual index
    %   d - Number of dimensions
    %
    % Output:
    %   solution - Generated deterministic solution

    solution = linspace(0, 1, d) * index / d;
end
```

```
function Pop = gridBasedInitialization(n, d, Xmin, Xmax)
    % Grid-Based Initialization
    % n: Population size (approximate total points)
    % d: Number of variables (dimensions)
    % Xmin: Lower bounds for each dimension
    % Xmax: Upper bounds for each dimension

    % Calculate the number of grid points per dimension
```

```
    grid_points_per_dimension = round(n^(1/d));

    % Generate grid vectors for each dimension
    grid_vectors = arrayfun(@(i) linspace(Xmin(i), Xmax(i),
     grid_points_per_dimension), 1:d, 'UniformOutput', false);

    % Create grid using ndgrid
    [grid_mesh{1:d}] = ndgrid(grid_vectors{:});

    % Reshape grid points into population matrix
    Pop = cell2mat(cellfun(@(x) x(:), grid_mesh, 'UniformOutput', false));
end
```

```
function Pop = sobol_sequence_initialization(n, d, Xmin, Xmax)
    % Sobol sequence initialization
    % Inputs:
    %   n    - Population size (number of individuals)
    %   d    - Number of variables (dimensions)
    %   Xmin - Lower bounds (1 x d vector)
    %   Xmax - Upper bounds (1 x d vector)
    % Output:
    %   Pop  - Initialized population (n x d matrix)

    % Create a Sobol sequence generator
    sobol_set = sobolset(d, 'Skip', 1e3, 'Leap', 10);
    sobol_set = scramble(sobol_set, 'MatousekAffineOwen');

    % Generate Sobol sequence points
    sobol_points = net(sobol_set, n);

    % Scale the points to the defined bounds
    Pop = Xmin + sobol_points .* (Xmax - Xmin);
end
```

4.2.9 *Oversampling Initialization*

Oversampling initialization generates numerous random solutions, selecting a quality-based subset [44], [45], to enhance the initial set and accelerate convergence. However, it involves high computational cost.

Steps:

Generate a large pool of candidate solutions using random or heuristic methods.

Select a subset based on diversity, fitness, or combined criteria for the initial population.

Characteristics

- Enhanced diversity: Generating extra solutions enables selection of candidates that maximize population diversity.
- Improved quality: Selection criteria can prioritize high-quality solutions, ensuring a stronger starting population.
- Higher computational cost: Oversampling demands more resources to evaluate a larger solution set upfront.

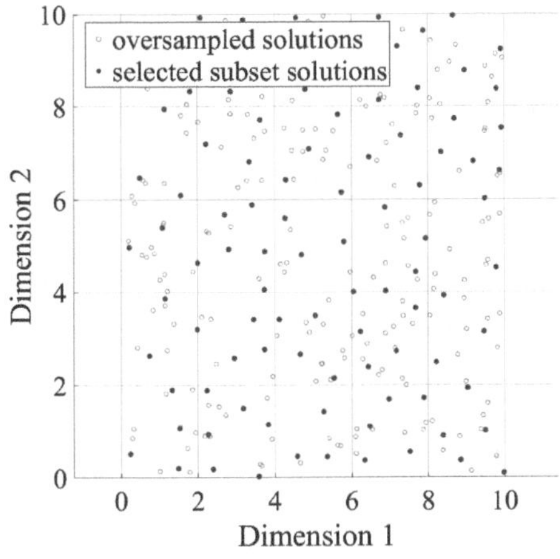

Figure 4.5: Oversampling initialization in 2D.

The initialization approach follows the outlined steps below.

Inputs:

n: Number of solutions in the final population.
m: Number of solutions generated during oversampling (where $m > n$).
GenerateCandidate(): Function to generate a solution.
SelectSubset(): Function to select n solutions from the oversampled set based on criteria such as fitness or diversity.

Outputs:

Pop: Final population of n solutions.

4.2.9.1 Pseudocode 1

```
BEGIN Oversampling Initialization

    # Define the problem
    Define the number of variables d
    Define the desired population size n
    Define the oversampling size m (m > n)

    # Oversampling: Generate a large set of candidate solutions
    Candidates = []
    FOR i from 1 to m
        solution_i = GenerateCandidate()  # Generate a candidate solution
        Add solution_i to Candidates
    END FOR

    # Selection: Choose n solutions from the Candidates set
    Pop = SelectSubset(Candidates, n)  # Select a subset based on fitness or
     diversity

    # Output the initialized population
    RETURN Pop

END Oversampling Initialization
```

Pseudocode explanation

- Oversampling: Generates m candidate solutions via **GenerateCandidate()**, using random, heuristic, or problem-specific methods.
- Selection: Selects n solutions from the oversampled set via **SelectSubset()**, prioritizing:
  - Fitness: Best-performing solutions based on the objective function.
  - Diversity: Ensuring broad search space coverage.
  - Hybrid criteria: Balancing fitness and diversity.
- Output: The final population Pop contains n solutions, selected for diversity, quality, or both.

4.2.9.2 Illustrative Example 1: Oversampling Initialization with Fitness-based Selection

- **Oversampling step**

4.2.9.3 Pseudocode 2

```
BEGIN Generate candidate solutions

# Generate a random solution within the problem bounds
solution = []
FOR each variable j in the solution vector
    solution[j] = Random_Value_Within_Bounds()
END FOR

RETURN solution

END Generate Candidate solutions
```

- **Selection Step**

4.2.9.4 Pseudocode 3

```
BEGIN Select subset of solutions

# Sort the candidates by their fitness (lower is better for
 minimization problems)
SortedCandidates = Sort(Candidates, by=Fitness)

# Select the top n solutions
Pop = []
FOR i from 1 to n
    Add SortedCandidates[i] to Pop
END FOR

RETURN Pop

END Select subset of solutions
```

4.2.9.5 Illustrative Example 2: Oversampling Initialization with Diversity-based Selection

- **Diversity-based selection step**

4.2.9.6 Pseudocode 4

```
BEGIN Select Subset by Diversity

# Initialize the selected population
Pop = []

# Add the first solution (arbitrarily) to the selected population
Add Candidates[1] to Pop
Remove Candidates[1] from Candidates

# Iteratively select solutions that maximize diversity
WHILE size of Pop < n
    best_candidate = None
    max_diversity = -Infinity

    FOR each candidate in Candidates
        diversity_score = CalculateDiversity(candidate, Pop)
        IF diversity_score > max_diversity
            best_candidate = candidate
            max_diversity = diversity_score
        END IF
    END FOR

    Add best_candidate to Pop
    Remove best_candidate from Candidates
END WHILE

RETURN Pop

END Select Subset by Diversity
```

- **Diversity calculation**

4.2.9.7 Pseudocode 5

```
BEGIN Calculate Diversity

# Compute the diversity of a candidate with respect to the selected
 population
diversity_score = 0
FOR each solution in Pop
    diversity_score += Distance(candidate, solution)
END FOR

RETURN diversity_score

END Calculate Diversity
```

4.2.9.8 Pros and Cons of Oversampling Initialization

Oversampling initialization generates a large candidate pool, selecting a subset based on specific criteria to enhance quality and diversity. While effective for optimization, it increases computational cost and complexity. See Table 4.5 for pros and cons.

Table 4.5: Comparison of pros and cons of oversampling initialization methods.

Pros	Cons
High-quality initialization: Evaluating a larger candidate set boosts the chance of starting with better solutions.	*Higher computational cost*: Generating and evaluating a larger solution set demands more time and resources.
Diversity preservation: Selection criteria can help cover the search space effectively, reducing the risk of premature convergence.	*Selection complexity*: Designing effective criteria is challenging, particularly in high-dimensional or multi-objective problems.
Flexibility: Enables customized selection criteria, balancing exploration and exploitation.	

4.2.9.9 *Example of Clustering Application*

In clustering problems [46], [47], oversampling generates many candidate cluster centers, selecting a subset based on data space coverage.

Example: K-means clustering initialization

4.2.9.10 *Pseudocode 6*

```
BEGIN Oversampling Initialization for K-Means

    # Oversampling: Generate m candidate cluster centers
    Candidates = []
    FOR i from 1 to m
        candidate = Random_Point_In_Data_Space()
        Add candidate to Candidates
    END FOR

    # Selection: Choose n cluster centers that maximize diversity
    Pop = SelectSubset(Candidates, n)  # Use diversity-based selection

    RETURN Pop

END Oversampling Initialization for K-Means
```

Concluding this section, we can state that oversampling initialization enhances quality and diversity in metaheuristics by generating a larger candidate set and applying careful selection criteria, establishing a strong optimization foundation. While computational costs increase, the benefits often justify the overhead, particularly in complex problems.

4.2.10 *MATLAB Code Examples of Oversampling Initialization*

```
function Pop = OversamplingInitialization(d, n, m)
    % Oversampling Initialization in MATLAB
    % d: Number of variables
    % n: Desired population size
    % m: Oversampling size (m > n)

    if m <= n
        error('Oversampling size m must be greater than n');
```

```
    end

    % Generate m candidate solutions
    Candidates = zeros(m, d);
    for i = 1:m
        Candidates(i, :) = GenerateCandidate(d);
    end

    % Select the best n solutions based on a selection criterion
    Pop = SelectSubset(Candidates, n);
end

function solution = GenerateCandidate(d)
    % Generate a random candidate solution within [0,1]^d (or modify as needed)
    solution = rand(1, d);
end

function subset = SelectSubset(Candidates, n)
    % Example selection: Select the first n solutions (modify as needed)
    % You can replace this with a fitness-based or diversity-based selection
    subset = Candidates(1:n, :);
end
```

```
function solution = GenerateCandidate(lowerBounds, upperBounds)
    % Generate a random solution within the problem bounds
    d = length(lowerBounds); % Number of variables
    solution = zeros(1, d);

    for j = 1:d
        solution(j) = lowerBounds(j) + (upperBounds(j) - lowerBounds(j)) * rand
    ();
    end
end
```

```
function Pop = SelectSubset(Candidates, FitnessValues, n)
    % Sort candidates by their fitness values in ascending order
    [~, sortedIndices] = sort(FitnessValues, 'ascend');
    SortedCandidates = Candidates(sortedIndices, :);

    % Select the top n solutions
    Pop = SortedCandidates(1:n, :);
end
```

```
function Pop = SelectSubsetByDiversity(Candidates, n)
    % SelectSubsetByDiversity selects a subset of n solutions maximizing
      diversity
    % Candidates: matrix of candidate solutions (rows are solutions, columns are
      variables)
    % n: desired number of selected solutions
    % Pop: selected subset of solutions

    % Initialize the selected population
    Pop = Candidates(1, :); % Add the first candidate arbitrarily
    Candidates(1, :) = []; % Remove the first candidate from the list

    while size(Pop, 1) < n
        best_candidate = [];
        max_diversity = -inf;

        for i = 1:size(Candidates, 1)
            candidate = Candidates(i, :);
            diversity_score = CalculateDiversity(candidate, Pop);

            if diversity_score > max_diversity
                best_candidate = candidate;
                max_diversity = diversity_score;
```

```
            end
        end

        % Add the best candidate to Pop and remove it from Candidates
        Pop = [Pop; best_candidate];
        Candidates(ismember(Candidates, best_candidate, 'rows'), :) = [];
    end
end

function score = CalculateDiversity(candidate, Pop)
    % CalculateDiversity computes the diversity of a candidate solution
    % relative to the existing population Pop using Euclidean distance

    if isempty(Pop)
        score = 0;
        return;
    end

    distances = sqrt(sum((Pop - candidate).^2, 2));
    score = min(distances); % Diversity is the minimum distance to existing
     solutions
end
```

```
function diversity_score = CalculateDiversity(candidate, Pop)
    % Compute the diversity of a candidate with respect to the selected
     population
    diversity_score = 0;

    for i = 1:size(Pop, 1)
        diversity_score = diversity_score + norm(candidate - Pop(i, :));
    end

    return;
end
```

```
function Pop = oversamplingInitializationKMeans(m, n, dataBounds)
    % Oversampling Initialization for K-Means
    % m: Number of candidate cluster centers
    % n: Desired number of selected cluster centers
    % dataBounds: [lower_bounds; upper_bounds] for each dimension

    d = size(dataBounds, 2); % Number of dimensions
    Candidates = zeros(m, d);

    % Step 1: Generate m candidate cluster centers
    for i = 1:m
        Candidates(i, :) = dataBounds(1, :) + rand(1, d) .* (dataBounds(2, :) -
     dataBounds(1, :));
    end

    % Step 2: Select n cluster centers maximizing diversity
    Pop = selectSubsetByDiversity(Candidates, n);
end

function Pop = selectSubsetByDiversity(Candidates, n)
    % Selects n points from Candidates maximizing diversity
    Pop = Candidates(1, :); % Start with the first candidate
    Candidates(1, :) = [];  % Remove from pool

    while size(Pop, 1) < n
        best_candidate = [];
        max_diversity = -inf;

        for i = 1:size(Candidates, 1)
            diversity_score = calculateDiversity(Candidates(i, :), Pop);
            if diversity_score > max_diversity
                best_candidate = Candidates(i, :);
```

```
                max_diversity = diversity_score;
            end
        end

        Pop = [Pop; best_candidate];
        Candidates(ismember(Candidates, best_candidate, 'rows'), :) = [];
    end
end

function diversity_score = calculateDiversity(candidate, Pop)
    % Computes diversity as the sum of distances to selected centers
    diversity_score = sum(vecnorm(Pop - candidate, 2, 2));
end
```

4.2.11 *Hybrid Initialization*

Hybrid initialization combines multiple techniques to balance random exploration and high-quality solutions from heuristic or knowledge-based methods [48], [49].

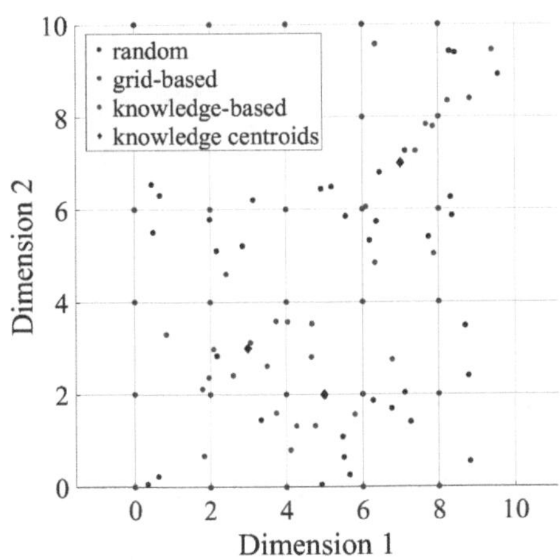

Figure 4.6: Hybrid initialization in 2D.

This strategy balances exploration and exploitation for greater flexibility. However, implementation is complex, requiring careful parameterization to ensure diversity and solution quality.

The hybrid approach typically involves:

1. Dividing the population into subsets, each generated using a different initialization method.
2. Combining the subsets into a single population for further processing.

This approach is especially useful in complex or multimodal optimization problems, where relying on a single method might lead to suboptimal solutions.

Steps for hybrid initialization

- **Define parameters**: Set population size and proportion for each initialization method.
- **Apply initialization**: Utilize random, heuristic-based, and deterministic methods to generate population subsets.
- **Merge subsets**: Combine all subsets into a single population for initialization.
- **Evaluate population**: Optionally assess initial population quality to refine optimization strategies.

Pseudocode example: An overview of the general method for implementing hybrid initialization.

4.2.11.1 Pseudocode 1

```
BEGIN HybridInitialization

    # Define parameters
    total_population_size = N
    proportion_random = p_random
    proportion_heuristic = p_heuristic
    proportion_deterministic = p_deterministic
    dimensions = d
    bounds = [lower_bound, upper_bound]

    # Calculate subset sizes
    random_size = floor(N * p_random)
    heuristic_size = floor(N * p_heuristic)
    deterministic_size = N - (random_size + heuristic_size)

    # Initialize subsets
    random_subset = []
    heuristic_subset = []
    deterministic_subset = []

    # Generate random subset
    FOR i = 1 TO random_size
        individual = GenerateRandomSolution(dimensions, bounds)
        random_subset.ADD(individual)
    END FOR

    # Generate heuristic subset
    FOR i = 1 TO heuristic_size
        individual = GenerateHeuristicSolution(dimensions, bounds)
        heuristic_subset.ADD(individual)
    END FOR

    # Generate deterministic subset
    FOR i = 1 TO deterministic_size
        individual = GenerateDeterministicSolution(dimensions, bounds)
        deterministic_subset.ADD(individual)
    END FOR

    # Merge subsets into a single population
    population = random_subset + heuristic_subset + deterministic_subset

    # Evaluate initial population (optional)
    fitness_values = []
    FOR individual IN population
        fitness = EvaluateFitness(individual)
```

```
        fitness_values.ADD(fitness)
    END FOR

    RETURN population, fitness_values

END HybridInitialization
```

Key Functions in the pseudocode

1. **GenerateRandomSolution(dimensions, bounds)**:
   Creates a solution by randomly sampling each dimension within its specified bounds.
2. **GenerateHeuristicSolution(dimensions, bounds)**:
   Uses problem-specific heuristics to exploit promising search space regions.
3. **GenerateDeterministicSolution(dimensions, bounds)**:
   Creates a solution using predefined rules or deterministic patterns, such as evenly spaced points.
4. **EvaluateFitness**:
   Computes a solution's fitness based on the objective function.

4.2.11.2 Pros and Cons of Hybrid Initialization

Hybrid initialization integrates random, heuristic, and deterministic methods to create a diverse, high-quality initial population. This flexibility allows algorithms to adapt to problem characteristics by combining multiple methods. However, higher complexity and computational costs present challenges. See Table 4.6 for pros and cons.

Optimizing hybrid initialization: Perform sensitivity analysis to determine optimal method ratios, use heuristic or lightweight deterministic techniques to minimize computational overhead, and experiment with problem-specific configurations. Hybrid initialization enhances metaheuristic algorithms by tackling these challenges.

4.2.12 *MATLAB Code Examples of Hybrid Initialization*

```
function [population, fitness_values] = HybridInitialization(N, p_random,
    p_heuristic, p_deterministic, dimensions, bounds)
    % Calculate subset sizes
    random_size = floor(N * p_random);
    heuristic_size = floor(N * p_heuristic);
    deterministic_size = N - (random_size + heuristic_size);

    % Initialize subsets
    random_subset = [];
    heuristic_subset = [];
    deterministic_subset = [];

    % Generate random subset
    for i = 1:random_size
        individual = GenerateRandomSolution(dimensions, bounds);
        random_subset = [random_subset; individual];
    end

    % Generate heuristic subset
```

Table 4.6: Comparison of Pros and cons of hybrid initialization methods.

Pros	Cons
Enhanced diversity: Combining methods ensures the population explores diverse areas of the solution space.	*Increased Complexity*: Combining multiple initialization methods increases implementation effort, making the algorithm more complex and harder to manage.
Better starting solutions: Heuristic and deterministic methods guide algorithms toward promising regions in the search space.	*Parameter tuning*: Carefully selecting population subset proportions (random, heuristic, deterministic) is crucial; improper tuning can hinder performance or negate hybridization benefits.
Adaptive flexibility: Method proportions can be adjusted to align with problem characteristics, optimizing performance.	*Higher computational cost*: Combining multiple initialization methods—especially heuristic and deterministic—increases computational overhead compared to simpler initialization approaches.
	Potential redundancy: Some methods may create overlapping solutions, reducing diversity despite their intended combination.
	Higher memory usage: Storing population subsets from different methods requires more memory, especially for large-scale problems.

```
    for i = 1:heuristic_size
        individual = GenerateHeuristicSolution(dimensions, bounds);
        heuristic_subset = [heuristic_subset; individual];
    end

    % Generate deterministic subset
    for i = 1:deterministic_size
        individual = GenerateDeterministicSolution(dimensions, bounds);
        deterministic_subset = [deterministic_subset; individual];
    end

    % Merge subsets into a single population
    population = [random_subset; heuristic_subset; deterministic_subset];

    % Evaluate initial population (optional)
    fitness_values = [];
    for i = 1:size(population, 1)
        fitness = EvaluateFitness(population(i, :));
        fitness_values = [fitness_values; fitness];
    end
end
```

4.3 Exercises to Practice all types of Initialization in MATLAB

4.3.1 *Practice Exercises for Random Initialization*

Here are two exercises to practice random initialization in MATLAB, focusing on its visualization and modification.

4.3.1.1 Exercise 1: Visualization of the Initial Population

Objective: Implement random initialization and plot the generated points.

Instructions:

- Implement random initialization in a two-dimensional space ($d = 2$) with limits $X_{\text{min}} = [0, 0]$ and $X_{\text{max}} = [10, 20]$.
- Generate a population of $n = 100$ random solutions.
- Use a `scatter` plot to visualize the distribution of the generated points.

Extra challenge:

- Experiment with different values of n and observe how the dispersion of points changes.

4.3.1.2 Exercise 2: Statistical Analysis of the Distribution of Solutions

Objective: Evaluate how the generated values are distributed in each dimension.

Instructions:

- Use $n = 500$ solutions and $d = 3$ dimensions with limits $X_{\text{min}} = [0, 0, 0]$ and $X_{\text{max}} = [10, 20, 30]$.
- Generate the random population.
- Calculate the means and standard deviations for each dimension.
- Generate box plots (`boxplot`) to analyze the distribution of values in each variable.

Extra challenge:

- Compare the generated distribution with a theoretical random uniform distribution using a histogram (`histogram`).

4.3.2 *Practice Exercises for Heuristic Initialization*

Next, 2 exercises are proposed to practice the heuristic initialization with a well-known optimization problem, such as the Traveling Salesman Problem (TSP), and an indicator to guide a heuristic such as diversity.

4.3.2.1 Exercise 1: Implementation of Heuristic Initialization in a Routing Problem

Objective: Implement a heuristic initialization algorithm for the Traveling Salesman Problem (TSP) using the nearest neighbor method.

Instructions:

- Define a set of 50 cities with random coordinates in a 2D space.
- Implement the nearest neighbor heuristic to generate an initial solution.
- Visualize the generated route using a line plot.

Extra challenge:

- Generate multiple heuristic initial solutions with different starting cities and compare their lengths.
- Overlay a scatter plot to show the distribution of the cities.

4.3.2.2 Exercise 2: Evaluation of Diversity in Heuristic Initialization

Objective: Analyze the diversity of solutions generated by a heuristic initialization compared to a random initialization.

Instructions:

- Define an optimization problem with $d = 2$ dimensions and limits $X_{\text{min}} = [0, 0]$ and $X_{\text{max}} = [50, 100]$.
- Generate two sets of initial solutions:
  - **Heuristic:** Use a simple heuristic (e.g., generate solutions near an estimated optimal point).
  - **Random:** Generate solutions completely at random.
- Compute the average distance between solutions within each set.
- Represent both distributions using scatter plots.

Extra challenge:

- Apply statistical metrics (mean and standard deviation of distances) to quantify the diversity of each method.
- Adjust the heuristic and observe how diversity changes in the generated set.

4.3.3 Practice Exercises for Knowledge-based Initialization

Proposed exercises: Explore knowledge-based initialization through vehicle routing and production scheduling optimization.

4.3.3.1 Exercise 1: Knowledge-based Initialization for Vehicle Routing Optimization

Objective: Utilize historical vehicle route data to generate more efficient initial solutions.

Instructions:

- Define a set of 100 customers with locations in a 2D space.
- Use historical route data to assign efficient initial routes.
- Generate an initial population of solutions based on adjusted historical routes.
- Visualize the generated initial routes using a plot.

Extra challenge:

- Compare the quality of the initial solutions generated with a random initialization method by measuring the total traveled distance in each case.

4.3.3.2 Exercise 2: Expert Knowledge-based Scheduling Optimization

Objective: Apply expert rules in generating initial solutions for a production scheduling problem.

Instructions:

- Define a set of 5 machines and 15 tasks with known processing times.
- Implement heuristic rules based on expert knowledge, such as:
  - Prioritizing tasks with shorter processing times first.
  - Grouping similar tasks on the same machine.
- Generate an initial population following these rules.
- Represent the initial production schedule graphically.

Extra challenge:

- Compare the diversity of the generated population with a random initialization method using statistical metrics such as the standard deviation of completion times.

4.3.4 *Practice Exercises for Deterministic Initialization*

Deterministic initialization: Practice grid-based initialization and low-discrepancy sequences (LDS) like the Sobol sequence.

4.3.4.1 Exercise 1: Grid-based Deterministic Initialization

Objective: Apply grid-based initialization to evenly distribute solutions in a 2D search space.

Instructions:

- Define a 2D search space with bounds $X_{\text{min}} = (-100, -100)$ and $X_{\text{max}} = (100, 100)$.
- Divide the space into a 30×30 grid.
- Generate an initial population of 900 solutions by placing them at grid points.
- Visualize the generated grid-based population using a scatter plot.

Extra challenge:

- Compare the uniformity of the grid-based initialization with a random initialization by computing the pairwise distances between points.

4.3.4.2 Exercise 2: Sobol Sequence-based Deterministic Initialization

Objective: Implement Sobol sequence-based initialization to achieve uniform coverage of a multidimensional search space.

Instructions:

- Define a 3D search space with bounds $X_{\text{min}} = (0, 0, 0)$ and $X_{\text{max}} = (1, 1, 1)$.
- Generate an initial population of 100 solutions using Sobol sequences.
- Scale the Sobol sequence points to fit within the defined bounds.
- Visualize the generated population using a 3D scatter plot.

Extra challenge:

- Compare the coverage efficiency of Sobol sequence initialization with random initialization by computing the average minimum distance between points.

4.3.5 *Practice Exercises for Oversampling Initialization*

4.3.5.1 Exercise 1: Fitness-based Oversampling Initialization

Oversampling in 2D space: A generic method enhances initialization by oversampling, using diversity indicators to assess candidate solution quality and sparsity.

Objective: Implement an oversampling initialization method that selects the best solutions based on a fitness criterion.

Instructions:

- Define a 2D search space with bounds $X_{min} = (-100, -100)$ and $X_{max} = (100, 100)$.
- Generate an initial set of $m = 200$ candidate solutions randomly distributed in the search space.
- Evaluate each candidate solution using a predefined fitness function (e.g., minimizing $f(x, y) = (x - 5)^2 + (y - 5)^2$).
- Select the best $n = 50$ solutions based on their fitness value.
- Visualize the initial candidate set and the selected subset using a scatter plot.

Extra challenge:

- Compare the performance of this initialization method with a purely random initialization by analyzing the mean fitness of the selected solutions.

4.3.5.2 Exercise 2: Diversity-based Oversampling Initialization

Objective: Implement an oversampling initialization method that selects the most diverse solutions based on pairwise distances.

Instructions:

- Define a 2D search space with bounds $X_{min} = (-30, -30)$ and $X_{max} = (30, 30)$.
- Generate an initial set of $m = 500$ candidate solutions randomly distributed in the search space.
- Select the most diverse $n = 100$ solutions by maximizing the minimum pairwise distance between selected solutions.
- Visualize the initial candidate set and the selected subset using a scatter plot.

Extra challenge:

- Compare the diversity of the selected subset with that of a purely random initialization by computing the average minimum distance between selected solutions.

4.3.6 *Practice Exercises for Hybrid Initialization*

To practice this type of initialization, different types of initializations already studied, such as grid, random or heuristic, are compared in order to make clear the purpose of this type of initialization.

4.3.6.1 Exercise 1: Hybrid Initialization with Fitness and Random Sampling

Objective: Implement a hybrid initialization method that combines random sampling with fitness-based selection to create a high-quality initial population.

Instructions:

- Define a 2D search space with bounds $X_{min} = (0, 0)$ and $X_{max} = (10, 10)$.
- Generate an initial set of $m = 50$ candidate solutions, where:
  - 70% of the solutions are randomly generated within the search space.
  - 30% of the solutions are generated by selecting the best solutions from an additional set of 30 randomly generated points based on the fitness function $f(x, y) = (x - 5)^2 + (y - 5)^2$.
- Merge the two subsets to form the final population of size m.
- Visualize the distribution of both subsets and the final population using a scatter plot.

Extra challenge:

- Compare the mean fitness of the hybrid-initialized population with a purely random initialization.

4.3.6.2 Exercise 2: Hybrid Initialization with Random, Grid-based, and Heuristic Sampling

Objective: Implement a hybrid initialization method that combines three approaches: random sampling, grid-based sampling, and heuristic-based initialization.

Instructions:

- Define a 2D search space with bounds $X_{min} = (0, 0)$ and $X_{max} = (10, 10)$.
- Generate an initial set of $m = 200$ candidate solutions using three methods:
  - 40% of the solutions are randomly generated.
  - 30% of the solutions are generated using a uniform grid over the search space.
  - 30% of the solutions are generated using a heuristic approach (e.g., clustering centers from a k-means algorithm applied to an additional set of randomly sampled points).

- Merge the three subsets to form the final population of size m.
- Visualize the initial distribution of each subset and the final merged population using a scatter plot.

Extra challenge:

- Compare the diversity of the hybrid-initialized population with a purely random initialization by computing the average minimum pairwise distance between solutions.

References

[1] N. Hansen, D. V. Arnold, and A. Auger, "Evolution strategies," in *Springer Handbook of Computational Intelligence*. Springer, 2015, pp. 871–898.

[2] J. Genlin, "Survey on genetic algorithm," *Computer Applications and Software*, vol. 2, no. 1, pp. 69–73, 2004.

[3] S. Das and P. N. Suganthan, "Differential evolution: A survey of the state-of-the-art," *IEEE Transactions on Evolutionary Computation*, vol. 15, no. 1, pp. 4–31, 2010.

[4] E. Rashedi, H. Nezamabadi-Pour, and S. Saryazdi, "Gsa: A gravitational search algorithm," *Information Sciences*, vol. 179, no. 13, pp. 2232–2248, 2009.

[5] R. Kundu, S. Chattopadhyay, S. Nag, M. A. Navarro, and D. Oliva, "Prism refraction search: A novel physics-based metaheuristic algorithm," *The Journal of Supercomputing*, vol. 80, no. 8, pp. 10 746–10 795, 2024.

[6] H. Salimi, "Stochastic fractal search: A powerful metaheuristic algorithm," *Knowledge-Based Systems*, vol. 75, pp. 1–18, 2015.

[7] L. A. Beltran, M. A. Navarro, D. Oliva, D. Campos-Peña, J. A. Ramos-Frutos, and S. Zapotecas-Martínez, "Quasi-random fractal search (qrfs): A dynamic metaheuristic with sigmoid population decrement for global optimization," *Expert Systems with Applications*, vol. 254, p. 124400, 2024.

[8] T. Zhang and Z. W. Geem, "Review of harmony search with respect to algorithm structure," *Swarm and Evolutionary Computation*, vol. 48, pp. 31–43, 2019.

[9] A. Faramarzi, M. Heidarinejad, B. Stephens, and S. Mirjalili, "Equilibrium optimizer: A novel optimization algorithm," *Knowledge-Based Systems*, vol. 191, p. 105190, 2020.

[10] P. Spettel, H.-G. Beyer, and M. Hellwig, "A covariance matrix self-adaptation evolution strategy for optimization under linear constraints," *IEEE Transactions on Evolutionary Computation*, vol. 23, no. 3, pp. 514–524, 2018.

[11] S. Mirjalili, "Sca: A sine cosine algorithm for solving optimization problems," *Knowledge-based Systems*, vol. 96, pp. 120–133, 2016.

[12] F. A. Hashim, K. Hussain, E. H. Houssein, M. S. Mabrouk, and W. Al-Atabany, "Archimedes optimization algorithm: a new metaheuristic algorithm for solving optimization problems," *Applied Intelligence*, vol. 51, no. 3, pp. 1531–1551, 2021.

[13] R. V. Rao and V. Patel, "An improved teaching-learning-based optimization algorithm for solving unconstrained optimization problems," *Scientia Iranica*, vol. 20, no. 3, pp. 710–720, 2013.

[14] M. A. Navarro, A. Casas-Ordaz, B. A. Rivera-Aguilar, B. Morales-Castañeda, and D. Oliva, "Population of hyperparametric solutions for the design of metaheuristic algorithms: An empirical analysis of performance in particle swarm optimization," in *Metaheuristics International Conference*. Springer, 2024, pp. 292–305.

[15] C. A. Coello Coello and R. L. Becerra, "Efficient evolutionary optimization through the use of a cultural algorithm," *Engineering Optimization*, vol. 36, no. 2, pp. 219–236, 2004.

[16] J. Kennedy and R. Eberhart, "Particle swarm optimization," in *Proceedings of ICNN'95-International Conference on Neural Networks*, vol. 4. IEEE, 1995, pp. 1942–1948.

[17] N. M. Al Salami, "Ant colony optimization algorithm," *UbiCC Journal*, vol. 4, no. 3, pp. 823–826, 2009.

[18] S. Mirjalili and A. Lewis, "The whale optimization algorithm," *Advances in Engineering Software*, vol. 95, pp. 51–67, 2016.

[19] B. Zolghadr-Asli, O. Bozorg-Haddad, and X. Chu, "Crow search algorithm (csa)," in *Advanced Optimization by Nature-inspired Algorithms*. Springer, 2018, pp. 143–149.

[20] O. Maciel, E. Cuevas, M. A. Navarro, D. Zaldívar, and S. Hinojosa, "Side-blotched lizard algorithm: a polymorphic population approach," *Applied Soft Computing*, vol. 88, p. 106039, 2020.

[21] A. A. Heidari, S. Mirjalili, H. Faris, I. Aljarah, M. Mafarja, and H. Chen, "Harris hawks optimization: Algorithm and applications," *Future Generation Computer Systems*, vol. 97, pp. 849–872, 2019.

[22] L. Abualigah, D. Yousri, M. Abd Elaziz, A. A. Ewees, M. A. Al-qaness, and A. H. Gandomi, "Aquila optimizer: A novel meta-heuristic optimization algorithm," *Computers & Industrial Engineering*, vol. 157, p. 107250, 2021.

[23] O. T. Altinoz, A. E. Yilmaz, and G.-W. Weber, "Improvement of the gravitational search algorithm by means of low-discrepancy sobol quasi random-number sequence based initialization," *Advances in Electrical and Computer Engineering*, vol. 14, no. 3, pp. 55–63, 2014.

[24] N. Q. Uy, N. X. Hoai, R. I. McKay, and P. M. Tuan, "Initialising pso with randomised low-discrepancy sequences: the comparative results," in *2007 IEEE Congress on Evolutionary Computation*. IEEE, 2007, pp. 1985–1992.

[25] A. Ashraf, S. Pervaiz, W. Haider Bangyal, K. Nisar, A. Ibrahim, A. Asri, D. B. Rawat *et al.*, "Studying the impact of initialization for population-based algorithms with low-discrepancy sequences," *Applied Sciences*, vol. 11, no. 17, p. 8190, 2021.

[26] T. Monica, A. Rajasekhar, M. Pant, and A. Abraham, "Enhancing the local exploration capabilities of artificial bee colony using low discrepancy sobol sequence," in *International Conference on Contemporary Computing*. Springer, 2011, pp. 158–168.

[27] P. Krömer, J. Platoš, and V. Snášel, "Differential evolution for the optimization of low-discrepancy generalized halton sequences," *Swarm and Evolutionary Computation*, vol. 54, p. 100649, 2020.

[28] P. A. Digehsara, S. N. Chegini, A. Bagheri, and M. P. Roknsaraei, "An improved particle swarm optimization based on the reinforcement of the population initialization phase by scrambled halton sequence," *Cogent Engineering*, vol. 7, no. 1, p. 1737383, 2020.

[29] S. J. Rosli, H. A. Rahim, K. N. Abdul Rani, R. Ngadiran, R. B. Ahmad, N. Z. Yahaya, M. Abdulmalek, M. Jusoh, M. N. M. Yasin, T. Sabapathy *et al.*, "A hybrid modified method of the sine cosine algorithm using latin hypercube sampling with the cuckoo search algorithm for optimization problems," *Electronics*, vol. 9, no. 11, p. 1786, 2020.

[30] W. D. Kelton, "Random initialization methods in simulation," *IIE Transactions*, vol. 21, no. 4, pp. 355–367, 1989.

[31] M. Abdel-Basset, L. Abdel-Fatah, and A. K. Sangaiah, "Metaheuristic algorithms: A comprehensive review," *Computational Intelligence for Multimedia Big Data on the Cloud with Engineering Applications*, pp. 185–231, 2018.

[32] S. A. Alsaidy, A. D. Abbood, and M. A. Sahib, "Heuristic initialization of pso task scheduling algorithm in cloud computing," *Journal of King Saud University-Computer and Information Sciences*, vol. 34, no. 6, pp. 2370–2382, 2022.

[33] M. Sam'an, M. Mat Deris, and Farikhin, "An improved grey wolf optimization based heuristic initialization algorithm for feature selection in p2p lending default prediction," *International Journal of Computers and Applications*, pp. 1–11, 2025.

[34] K. L. Hoffman, M. Padberg, G. Rinaldi *et al.*, "Traveling salesman problem," *Encyclopedia of Operations Research and Management Science*, vol. 1, pp. 1573–1578, 2013.

[35] C. Li, X. Chu, Y. Chen, and L. Xing, "A knowledge-based technique for initializing a genetic algorithm," *Journal of Intelligent & Fuzzy Systems*, vol. 31, no. 2, pp. 1145–1152, 2016.

[36] E. P. Klement, E. Lughofer, J. Himmelbauer, and B. Moser, *Data-driven and Knowledge-based Modeling*. Springer, 2009.

[37] F. Alonso, L. Martínez, A. Pérez, and J. P. Valente, "Cooperation between expert knowledge and data mining discovered knowledge: Lessons learned," *Expert Systems with Applications*, vol. 39, no. 8, pp. 7524–7535, 2012.

[38] T. Feng, J. Li, H. Jiang, S. X. Yang, P. Wang, Y. Teng, S. Chen, Q. Fu, and B. Luo, "The optimal global path planning of mobile robot based on improved hybrid adaptive genetic algorithm in different tasks and complex road environments," *IEEE Access*, vol. 12, pp. 18 400–18 415, 2024.

[39] S. Roychowdhury, M. Diligenti, and M. Gori, "Regularizing deep networks with prior knowledge: A constraint-based approach," *Knowledge-Based Systems*, vol. 222, p. 106989, 2021.

[40] A. Borghesi, F. Baldo, and M. Milano, "Improving deep learning models via constraint-based domain knowledge: A brief survey," *arXiv preprint arXiv:2005.10691*, 2020.

[41] S. Elsayed, R. Sarker, and C. A. C. Coello, "Sequence-based deterministic initialization for evolutionary algorithms," *IEEE Transactions on Cybernetics*, vol. 47, no. 9, pp. 2911–2923, 2016.

[42] M. D. McKay, R. J. Beckman, and W. J. Conover, "A comparison of three methods for selecting values of input variables in the analysis of output from a computer code," *Technometrics*, vol. 42, no. 1, pp. 55–61, 2000.

[43] I. M. Sobol', D. Asotsky, A. Kreinin, and S. Kucherenko, "Construction and comparison of high-dimensional sobol' generators," *Wilmott*, vol. 2011, no. 56, pp. 64–79, 2011.

[44] F. Fernández-Navarro, C. Hervás-Martínez, and P. A. Gutiérrez, "A dynamic over-sampling procedure based on sensitivity for multi-class problems," *Pattern Recognition*, vol. 44, no. 8, pp. 1821–1833, 2011.

[45] S. Sharma, A. Gosain, and S. Jain, "A review of the oversampling techniques in class imbalance problem," in *International Conference on Innovative Computing and Communications: Proceedings of ICICC 2021, Volume 1*. Springer, 2022, pp. 459–472.

[46] J. Zhang and L. Chen, "Clustering-based undersampling with random over sampling examples and support vector machine for imbalanced classification

of breast cancer diagnosis," *Computer Assisted Surgery*, vol. 24, no. sup2, pp. 62–72, 2019.

[47] Z. Xu, D. Shen, T. Nie, Y. Kou, N. Yin, and X. Han, "A cluster-based oversampling algorithm combining smote and k-means for imbalanced medical data," *Information Sciences*, vol. 572, pp. 574–589, 2021.

[48] G. S. Martínez, T. Karhela, H. Niemistö, A. Rossi, C. Pang, and V. Vyatkin, "A hybrid approach for the initialization of tracking simulation systems," in *2015 IEEE 20th Conference on Emerging Technologies & Factory Automation (ETFA)*. IEEE, 2015, pp. 1–8.

[49] Y. Jiang, C. Ma, S. Wang, and Y. Li, "A novel evolutionary algorithm for structural model updating with a hybrid initialization and multi-stage update strategy," *Ocean Engineering*, vol. 305, p. 117836, 2024.

Chapter 5

Implementation of Initialization Methods in PSO and DE

5.1 Introduction

A crucial step for PSO and DE, influencing convergence, solution quality, and avoidance of local optima. PSO mimics swarm behavior for guided search, while DE uses individual differences for exploration. This chapter covers random, Latin Hypercube Sampling, Opposition-Based Learning, and Chaos-Based initialization, analyzing their impact on exploration, exploitation, and diversity. It also provides implementation guidance, pseudocode, and trade-offs, equipping readers to enhance metaheuristic algorithms for complex optimization problems.

5.1.1 *Background*

DE application distribution: Fig. 5.1 highlights Differential Evolution (DE) usage across fields in Scopus. It is most prominent in Computer Science (29.0%), aiding AI, machine learning, and hyperparameter optimization. Engineering (25.9%) follows, supporting system design and industrial automation. Mathematics (15.0%) benefits from its computational foundation. Other domains include Energy (4.7%), Materials (4.0%), Physics & Astronomy (4.0%), and Decision Sciences (3.4%), showing its relevance in optimization-driven applications.

DOI: 10.1201/9781032723563-5

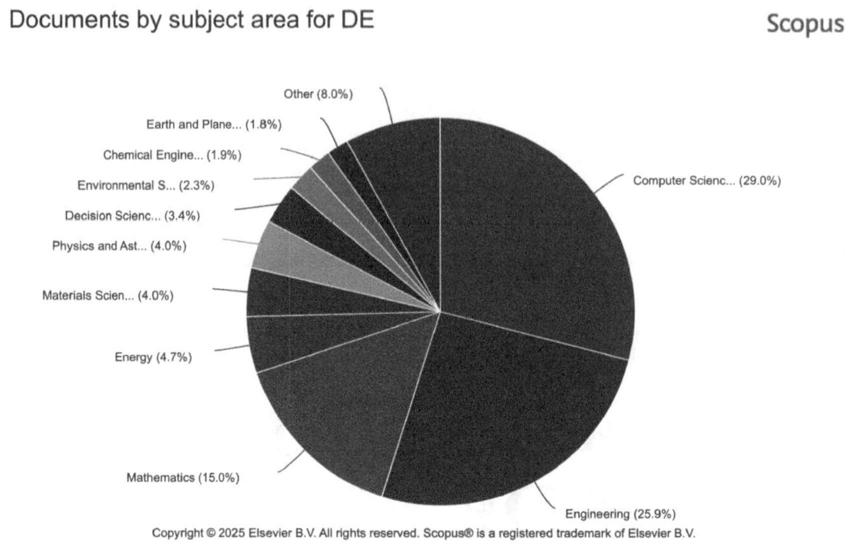

Figure 5.1: Documents by subject area for DE.

PSO application distribution: Fig. 5.2 shows Particle Swarm Optimization (PSO) usage across subject areas in Scopus. It is most prominent in Engineering (28.5%), supporting system design, automation, and optimization of mechanical and electrical structures. Computer Science (21.5%) follows, aiding AI, machine learning, and neural network optimization. Mathematics (10.1%) leverages PSO for numerical optimization. Other fields include Energy (6.7%), Physics & Astronomy (5.8%), Materials Science (5.3%), Environmental Sciences (3.7%), Chemistry (2.1%), and Earth & Planetary Sciences (2.4%), demonstrating its diverse applications in computational and scientific research.

Differential Evolution (DE) and Particle Swarm Optimization (PSO) are key metaheuristic algorithms widely applied in Engineering, Computer Science, and Mathematics. DE specializes in numerical and combinatorial optimization, aiding engineering design, machine learning, and multi-objective problems. PSO thrives in high-dimensional search spaces, benefiting AI, control systems, and energy management. This book examines their theoretical foundations, implementations, and optimization strategies, focusing on initialization methods to enhance search diversity and convergence efficiency.

5.1.1.1 Particle Swarm Optimization

Introduced in 1995 by Russell C. Eberhart and James Kennedy, PSO is inspired by the social behavior of bird flocks, modeling how they forage for food and share information. While PSO shares characteristics with evolutionary algorithms

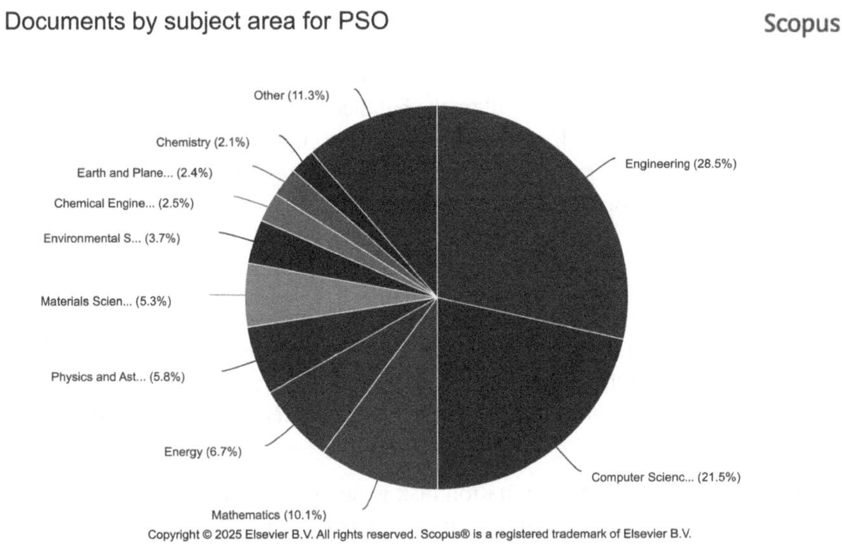

Figure 5.2: Documents by subject area for PSO.

(EAs) [1] and genetic algorithms (GAs) [2], it differs significantly in structure and simplicity.

Candidate solution initialization is crucial in population-based metaheuristics (MAs), typically using pseudorandom number generators (PRNGs) to sample the first generation. In PSO, the initialization equation:

$$x_k^{i,t} = l_k + rand(u_k - l_k), \qquad x_k^{i,t} \in x^t \tag{5.1}$$

Where, $x_k^{i,t}$ is the ith population particle x^l, i represents the number of particles and its size is $(i = 1, 2, 3, ..., N_{particles})$, k defines the problem dimension, the iteration number is represented by t, meanwhile l_k and u_k defines a particle's position within upper u_k and lower l_k limits, where $rand$ is a random value between 0 and 1. Once positions are established, velocity updates direct particle movement across the search space.

5.1.1.2 Velocity and Movement of Particles

To find the velocity, for each solution it is necessary to compute the positions of the global (g_{best}) and local optimal (l_{best}). Equation 5.2 defines the velocity calculation for each particle.

$$v^{t+1} = w \times v^t + c_1 \times rand1 \times (l_{best} - x^t) + c_2 \times rand2 \times (g_{best} - x^t) \tag{5.2}$$

$$x^{t+1} = x^t + v^{t+1} \tag{5.3}$$

From Equations 5.2 and 5.3, v^{t+1} represents the velocity at iteration $t+1$, while v^t denotes the velocity of the previous iteration. The particle positions are stored in the vector x^t. Two random variables, $rand1$ and $rand2$, are drawn from a uniform distribution $(0, 1)$. Constants $c_{1,2}$ regulate the balance between local and global optima. During the current iteration, Equation 5.3 is used to update particle particles. The new positions are stored in x^{t+1}, while x^t contains the previous positions. Finally, the velocity vector v^{t+1} is computed using Equation 5.2.

5.1.1.3 Differential Evolution

Introduced by Storn and Price [3], DE optimizes real parameters without requiring gradient operations. It initializes candidate solutions using a uniform distribution within search space limits. Given $x \in R^D$, a candidate solution in a D-dimensional space, the problem is formulated as $f : R_D \longrightarrow R$, typically aiming for minimization.

During DE's iterative process, a mutation task generates a vector v_i for each x_i. This mutation step is applied in each iteration G across the entire population producing vectors v_i. The most common mutation strategy, *DE/rand/1*, is defined as:

$$DE/rand/1 \longrightarrow v_i = x_{r1} + F(x_{r2} - x_{r3}) \tag{5.4}$$

From Equation 5.4, r_1, r_2, r_3 are distinct random integers ($r_1 \neq r_2 \neq r_3 \neq i$) drawn from the range $[1, NP]$, where NP denotes the population size. These indices are generated per mutant vector. The parameter F is a scaling factor, acting as a control variable to adjust the influence of the difference vector.

Over the years, DE has gained popularity as a methodology for solving complex optimization problems due to its simplicity and effectiveness, requiring only a few parameters to influence performance. The standard DE version relies on three key parameters:

Population size (NP) must satisfy $NP \geq 4$. Mutation factor (F), also known as differential weight, is selected within $F \in [0, 2]$. Crossover probability (CR) governs the crossover operator, chosen from $CR \in [0, 1]$. Common parameter settings include: $F = 1.0, CR = 0.1$; $F = 1.0, CR = 0.9$; $F = 0.8, CR = 0.2$.

5.2 Test Initializations Considered for Application

5.2.1 *Latin Hypercube Sampling (LHS)*

Latin Hypercube Sampling (LHS) is a statistical method used to generate sample points from a multidimensional distribution while ensuring that each variable is well represented across its range. Originally, introduced by McKay, Beckman, and Conover in 1979 [4], LHS is widely employed in uncertainty quantification, optimization, and surrogate modeling due to its ability to provide better coverage of the search space compared to purely random sampling.

5.2.1.1 Fundamental Concepts

LHS is a stratified sampling technique ensuring uniform coverage across each variable's range, outperforming traditional Monte Carlo methods. In a d-dimensional space, LHS divides dimensions into n non-overlapping intervals, selecting one sample per interval, ensuring complete representation.

5.2.1.2 Mathematical Definition

Consider a d-dimensional space where we seek to generate a sample set $X = x_1, x_2, \ldots, x_n$ of n points. The LHS procedure follows these steps:

Divide the range of each variable X_j into n equal probability intervals:

$$I_{j,i} = \left[\frac{i-1}{n}, \frac{i}{n}\right], \quad i = 1, \ldots, n, \quad j = 1, \ldots, d. \tag{5.5}$$

For each dimension j, randomly select a point $u_{j,i}$ from each interval $I_{j,i}$:

$$u_{j,i} \sim \text{Uniform}\left(\frac{i-1}{n}, \frac{i}{n}\right). \tag{5.6}$$

Permute the sampled points within each dimension independently to avoid clustering effects.

Construct the final sample set by mapping the scaled values to the target distribution using an inverse cumulative distribution function (CDF) F_j^{-1}:

$$x_{j,i} = F_j^{-1}(u_{j,i}), \quad j = 1, \ldots, d, \quad i = 1, \ldots, n. \tag{5.7}$$

5.2.1.3 Advantages and Applications

Latin Hypercube Sampling (LHS) improves space-filling by avoiding point clustering, ensuring better parameter representation and reducing estimation variance for stable statistical analysis. It supports varied probability distributions, making it versatile for applications like:

- Engineering optimization: Enhancing surrogate modeling for efficient design space exploration.
- Computational simulations: Crucial for sensitivity analysis and capturing parameter variations.
- Machine learning: Used in hyperparameter optimization for systematic configuration searches.
- Environmental & reliability assessments: Helps in uncertainty analysis and solution optimization.

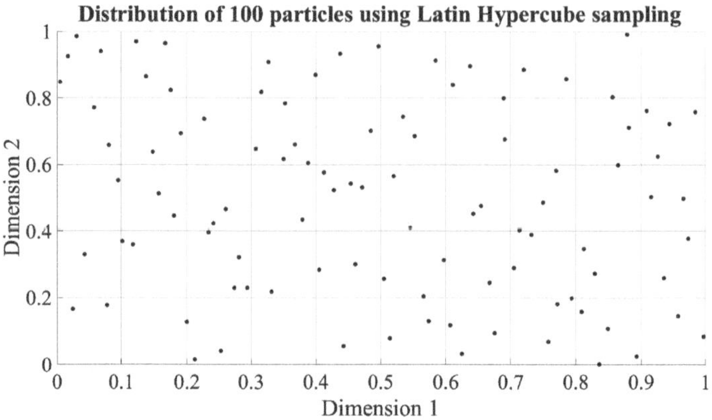

Figure 5.3: 100 solutions candidates using Latin Hypercube sampling.

5.2.1.3.1 Pseudocode 1

```
BEGIN Latin Hypercube Sampling Initialization

  # Define the problem
  Define the number of dimensions d
  Define the number of particles numParticles
  Define the lower and upper bounds for each dimension

  # Generate Latin Hypercube Sampling (LHS) matrix
  LHS_Matrix = LatinHypercubeSampling(numParticles, d)

  # Scale the sampled points to the defined limits
  FOR each particle i in LHS_Matrix
      ScaledSolution_i = lowerBounds + (LHS_Matrix[i] * (upperBounds -
   lowerBounds))
      Add ScaledSolution_i to Population
  END FOR

  # Return the initialized population
  RETURN Population

END Latin Hypercube Sampling Initialization
```

5.2.2 *Opposition-Based Learning (OBL)*

Opposition-Based Learning (OBL), introduced by Tizhoosh (2005) [5], OBL enhances optimization algorithms by evaluating candidate solutions alongside their opposites. Widely applied in evolutionary algorithms and machine learning, OBL improves exploration, exploitation, and convergence speed, leading to higher-quality solutions.

5.2.2.1 Fundamental Concepts

The core concept of OBL is that if a candidate solution is suboptimal, its opposite counterpart has a higher chance of being closer to the optimal. This approach enhances diversity, especially in high-dimensional search spaces, where random initialization may fall short. OBL is commonly applied in initialization and generation jumping to improve sampling of promising regions in optimization algorithms.

5.2.2.2 Mathematical Definition

For a solution space defined by an interval $X_j \in [a_j, b_j]$ in each dimension $j = 1, \ldots, d$, a candidate solution $X = (x_1, x_2, \ldots, x_d)$ is complemented by its opposite solution $\hat{X} = (\hat{x}_1, \hat{x}_2, \ldots, \hat{x}_d)$, where each component is computed as:

$$\hat{x}_j = a_j + b_j - x_j, \quad j = 1, \ldots, d. \tag{5.8}$$

This transformation ensures that if X is randomly sampled within its bounds, its opposite counterpart $\hat{X}$ remains a valid sample within the same domain, ensuring balanced exploration of the search space.

For normalized search spaces where $X_j \in [0, 1]$, the opposite solution simplifies to:

$$\hat{x}_j = 1 - x_j, \quad j = 1, \ldots, d. \tag{5.9}$$

The opposition-based concept can be further extended to quasi-oppositional learning, where instead of directly using $\hat{X}$, a weighted balance between the original and opposite solutions is considered to enhance convergence:

$$\tilde{x}_j = \alpha x_j + (1 - \alpha)\hat{x}_j, \quad \alpha \in [0, 1]. \tag{5.10}$$

5.2.2.3 Advantages and Applications

Opposition-Based Learning (OBL) accelerates convergence by reducing iteration count and enhancing search space exploration. In evolutionary computing, it improves initialization diversity, increasing proximity to optimal solutions. Integrated into Differential Evolution, PSO, and Genetic Algorithms, OBL boosts performance and convergence. In machine learning, it refines neural network training, feature selection, and reinforcement learning, improving model generalization through enhanced sampling strategies.

5.2.2.3.1 Pseudocode 2

```
BEGIN Opposition-Based Learning Initialization

    # Define the problem
    Define the number of dimensions d
    Define the number of particles numParticles
    Define the lower and upper bounds for each dimension

    # Generate initial random particles
    FOR each particle i in numParticles
        Particle_i = lowerBounds + rand() * (upperBounds - lowerBounds)
        Add Particle_i to Population
    END FOR
```

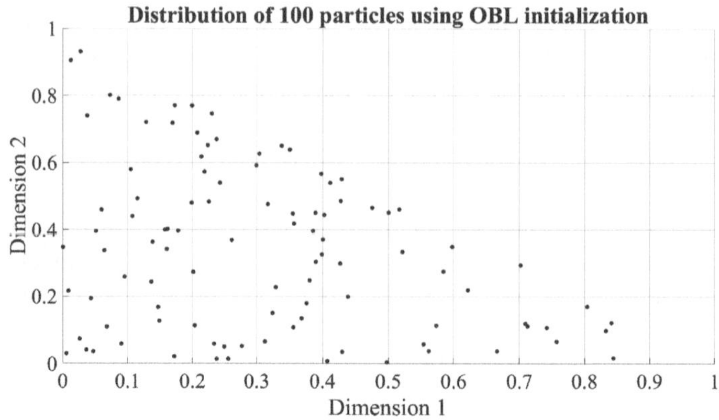

Figure 5.4: 100 solution candidates using Opposition-Based Learning sampling.

```
    # Generate opposite particles
    FOR each particle i in Population
        OppositeParticle_i = lowerBounds + (upperBounds - Particle_i)
        Add OppositeParticle_i to OppositePopulation
    END FOR

    # Combine original and opposite particles
    AllParticles = Population \cup OppositePopulation

    # Evaluate all particles using a fitness function
    FOR each particle i in AllParticles
        Fitness_i = ObjectiveFunction(AllParticles[i])
    END FOR

    # Select the best particles based on fitness values
    Sort AllParticles by Fitness in ascending order
    SelectedParticles = first numParticles from AllParticles

    # Return the selected initial population
    RETURN SelectedParticles

END Opposition-Based Learning Initialization
```

5.2.3 *Chaos-Based Sampling*

Chaos-Based Sampling is an initialization method utilizing chaotic system properties to generate diverse and structured sample points [6]. Unlike random sampling, it employs deterministic nonlinear maps for complex yet ordered sequences. Applied in optimization, metaheuristics, and machine learning, it improves exploration and prevents premature convergence.

5.2.3.1 Fundamental Concepts

Chaos theory examines dynamic systems with sensitivity to initial conditions, producing deterministic yet seemingly random behavior [7], [8]. In sampling, chaotic sequences enable structured randomness, enhancing diversity and coverage beyond stochastic methods. Its key advantage is balancing exploration and exploitation, as chaotic maps generate densely distributed points without clustering.

5.2.3.2 Mathematical Definition

A chaos-based sequence can be generated utilizing chaotic maps like the logistic, tent, and sine maps generating iterative, deterministic sequences with complex behavior, enhancing diversity in search spaces.

One of the most commonly used chaotic maps is the Logistic Map, defined as:

$$x_{n+1} = rx_n(1 - x_n), \quad x_n \in (0, 1), \quad r \in (0, 4), \tag{5.11}$$

where r is the control parameter that determines the chaotic behavior of the sequence. For $r \geq 3.57$, the sequence enters a chaotic regime, making it suitable for sampling applications.

Similarly, the tent map is another widely used chaotic function, given by:

$$x_{n+1} = \begin{cases} \frac{x_n}{\mu}, & \text{if } x_n < \mu, \\ \frac{1-x_n}{1-\mu}, & \text{if } x_n \geq \mu. \end{cases} \tag{5.12}$$

where μ is a parameter controlling the shape of the map.

To generate a sample set $X = x_1, x_2, \ldots, x_n$ using chaos-based sampling, an initial value x_0 is chosen, and the selected chaotic map is iterated to produce a sequence. The generated values are then scaled to match the desired range of the optimization domain:

$$x'_i = a + (b - a)x_i, \quad i = 1, \ldots, n, \tag{5.13}$$

where $[a, b]$ defines the problem's search space bounds.

5.2.3.3 Advantages and Applications

Chaos-based sampling enhances diversity compared to traditional random sampling while maintaining a structured distribution of points. This enhances the performance of optimization algorithms by preventing early stagnation and ensuring thorough exploration of the search space. Applications include evolutionary algorithms, swarm intelligence techniques, and neural network training, where effective initialization is crucial for convergence and efficiency. Additionally, chaos-based methods have been employed in cryptographic key generation, signal processing, and financial modeling due to their pseudo-random yet reproducible properties.

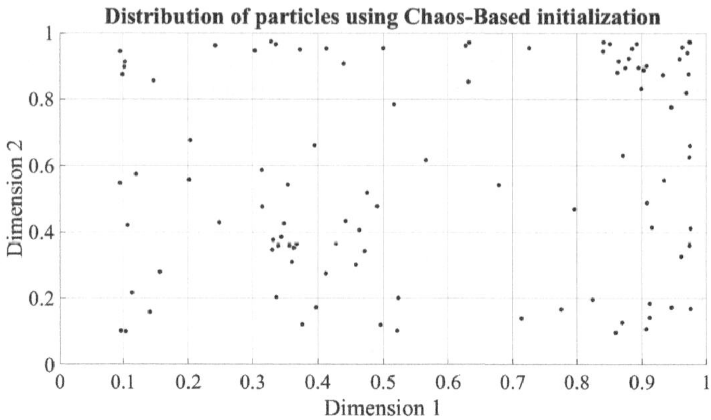

Figure 5.5: 100 solutions candidates using Chaos-Based sampling.

5.2.3.3.1 Pseudocode 3

```
BEGIN Chaos-Based Initialization

    # Define the problem
    Define the number of variables d
    Define the search space bounds lowerBounds and upperBounds

    # Define parameters for the logistic map
    chaosIterations = numParticles * d  # Total chaotic values required
    r = 3.9  # Logistic map parameter in the chaotic range
    x0 = 0.5  # Initial seed value

    # Generate chaotic sequence
    Initialize chaosSequence with size chaosIterations
    chaosSequence[1] = x0
    FOR i FROM 2 TO chaosIterations
        chaosSequence[i] = r * chaosSequence[i-1] * (1 - chaosSequence[i-1])
    END FOR

    # Remap chaotic sequence to the range [0,1]
    Apply mod function to chaosSequence

    # Reshape and scale to search space
    Reshape chaosSequence into a matrix of size (numParticles, d)
    Scale chaosMatrix within lowerBounds and upperBounds
    Store results in particles

    # Return the chaos-initialized population
    RETURN particles

END Chaos-Based Initialization
```

5.2.3.4 MATLAB® Code for Showing Initializations Considered for Application

```
% Configuration
numParticles = 100; % Number of particles
dim = 2; % Dimension of the search space (2D)
lowerBounds = [0, 0]; % Lower bound for each dimension
```

```
upperBounds = [1, 1]; % Upper bound for each dimension

% Initialization using Latin Hypercube Sampling
lhsMatrix = lhsdesign(numParticles, dim); % LHS matrix (values between 0 and 1)

% Scaling to the defined limits
particles = bsxfun(@plus, lowerBounds, bsxfun(@times, lhsMatrix, (upperBounds -
    lowerBounds)));

% Plot
figure;
scatter(particles(:,1), particles(:,2), 50, 'filled', 'MarkerFaceColor', 'b');
title('Distribution of 100 particles using Latin Hypercube Sampling');
xlabel('Dimension 1');
ylabel('Dimension 2');
grid on;
axis([lowerBounds(1) upperBounds(1) lowerBounds(2) upperBounds(2)]);
```

```
% Configuration
numParticles = 100; % Number of particles
dim = 2; % Search space dimension (2D)
lowerBounds = [0, 0]; % Lower bound for each dimension
upperBounds = [1, 1]; % Upper bound for each dimension

% Generate initial random particles
particles = lowerBounds + rand(numParticles, dim) .* (upperBounds - lowerBounds)
    ;

% Generate opposite particles
oppositeParticles = lowerBounds + (upperBounds - particles);

% Combine original and opposite particles
allParticles = [particles; oppositeParticles];

% Dummy evaluation to select the best particles (example: sum of coordinates)
fitness = sum(allParticles, 2); % Here you can use your objective function
[~, bestIndices] = mink(fitness, numParticles); % Select the best particles

% Select the initial set based on the best particles
selectedParticles = allParticles(bestIndices, :);

% Plot
figure;
scatter(selectedParticles(:, 1), selectedParticles(:, 2), 50, 'filled', '
    MarkerFaceColor', 'b');
title('Particle distribution using Opposition-Based Learning');
xlabel('Dimension 1');
ylabel('Dimension 2');
grid on;
axis([lowerBounds(1) upperBounds(1) lowerBounds(2) upperBounds(2)]);
```

```
% Configuration
numParticles = 100; % Number of particles
dim = 2; % Dimension of the search space (2D)
lowerBounds = [0, 0]; % Lower bound for each dimension
upperBounds = [1, 1]; % Upper bound for each dimension

% Logistic map parameters
chaosIterations = numParticles * dim; % Total chaotic values required
r = 3.9; % Logistic map parameter (chaotic range)
x0 = 0.5; % Initial seed value

% Generate chaotic sequence
chaosSequence = zeros(chaosIterations, 1);
chaosSequence(1) = x0;
for i = 2:chaosIterations
    chaosSequence(i) = r * chaosSequence(i-1) * (1 - chaosSequence(i-1));
```

```
end

% Remap chaotic sequence to the range [0, 1]
chaosSequence = mod(chaosSequence, 1);

% Reshape and scale to the search space
chaosMatrix = reshape(chaosSequence, [numParticles, dim]);
particles = lowerBounds + chaosMatrix .* (upperBounds - lowerBounds);

% Plot
figure;
scatter(particles(:, 1), particles(:, 2), 50, 'filled', 'MarkerFaceColor', 'b');
title('Particle Distribution using Chaos-Based Initialization');
xlabel('Dimension 1');
ylabel('Dimension 2');
grid on;
axis([lowerBounds(1) upperBounds(1) lowerBounds(2) upperBounds(2)]);
```

5.3 Initializations Applied to PSO and DE

The three initialization methods—Latin Hypercube Sampling (LHS), Opposition-Based Learning (OBL), and Chaos-Based Sampling—are widely studied and applied across various optimization contexts. In this subsection, they are implemented within the foundational structures of two canonical metaheuristic algorithms, Differential Evolution (DE) and Particle Swarm Optimization (PSO), resulting in six algorithm-initialization combinations, as outlined below.

- PSO-LHS
- PSO-OBL
- PSO-CHAOS
- DE-LHS
- DE-OBL
- DE-CHAOS

To assess the performance of the implemented algorithms, PSO and DE were coded to optimize the Ackley function (see Appendix A), a challenging multimodal function with numerous suboptimal solutions, complicating optimization. Below are MATLAB codes demonstrating PSO with three different initialization methods applied to optimize a test function.

The following code defines a function designed to combine PSO-LHS and optimize the search space of the Ackley function.

```
function pso_ackley_lhs()
    % PSO Parameters
    num_particles = 30; % Number of particles
    num_dimensions = 2; % Problem dimensions
    max_iterations = 100; % Maximum iterations
    w = 0.7; % Inertia weight
    c1 = 1.5; % Cognitive coefficient
    c2 = 1.5; % Social coefficient
    bounds = [-32.768, 32.768]; % Ackley function bounds
```

```
    % Initialization with Latin Hypercube Sampling
    X = lhsdesign(num_particles, num_dimensions) * (bounds(2) - bounds(1)) +
     bounds(1);
    V = zeros(num_particles, num_dimensions); % Initial velocities

    % Initial evaluation
    fitness = ackley(X);
    pbest = X; % Best individual positions
    pbest_fitness = fitness;

    % Global best
    [gbest_fitness, idx] = min(pbest_fitness);
    gbest = pbest(idx, :);

    % PSO iterations
    for iter = 1:max_iterations
        % Update velocity and position
        r1 = rand(num_particles, num_dimensions);
        r2 = rand(num_particles, num_dimensions);
        V = w * V + c1 * r1 .* (pbest - X) + c2 * r2 .* (gbest - X);
        X = X + V;

        % Apply search space constraints
        X = max(min(X, bounds(2)), bounds(1));

        % Evaluate objective function
        fitness = ackley(X);

        % Update best individual positions
        update_mask = fitness < pbest_fitness;
        pbest(update_mask, :) = X(update_mask, :);
        pbest_fitness(update_mask) = fitness(update_mask);

        % Update global best
        [new_gbest_fitness, idx] = min(pbest_fitness);
        if new_gbest_fitness < gbest_fitness
            gbest_fitness = new_gbest_fitness;
            gbest = pbest(idx, :);
        end

        % Display progress
        fprintf('Iteration %d: Best value found = %.5f\n', iter, gbest_fitness);
    end

    fprintf('Best solution found: [%.5f, %.5f]\n', gbest(1), gbest(2));
end

function f = ackley(X)
    a = 20;
    b = 0.2;
    c = 2 * pi;

    n = size(X, 2);
    sum1 = sum(X.^2, 2);
    sum2 = sum(cos(c * X), 2);

    f = -a * exp(-b * sqrt(sum1 / n)) - exp(sum2 / n) + a + exp(1);
end
```

The following code shows the PSO-OBL algorithm denoting the second initialization method for PSO.

```
function pso_ackley_obl()
    % PSO Parameters
    num_particles = 30;  % Number of particles
    num_dimensions = 2;  % Problem dimensions
    max_iterations = 100; % Maximum iterations
    w = 0.7;   % Inertia weight
    c1 = 1.5;  % Cognitive coefficient
    c2 = 1.5;  % Social coefficient
    bounds = [-32.768, 32.768]; % Ackley function bounds

    % Initialization with Opposition-Based Learning
    X = rand(num_particles, num_dimensions) * (bounds(2) - bounds(1)) + bounds
    (1);
    X_opposite = bounds(1) + bounds(2) - X; % Opposition-based points

    % Evaluate both original and opposite solutions
    fitness_X = ackley(X);
    fitness_X_opposite = ackley(X_opposite);

    % Select the better population
    mask = fitness_X_opposite < fitness_X;
    X(mask, :) = X_opposite(mask, :);

    V = zeros(num_particles, num_dimensions); % Initial velocities

    % Initial evaluation
    fitness = ackley(X);
    pbest = X; % Best individual positions
    pbest_fitness = fitness;

    % Global best
    [gbest_fitness, idx] = min(pbest_fitness);
    gbest = pbest(idx, :);

    % PSO iterations
    for iter = 1:max_iterations
        % Update velocity and position
        r1 = rand(num_particles, num_dimensions);
        r2 = rand(num_particles, num_dimensions);
        V = w * V + c1 * r1 .* (pbest - X) + c2 * r2 .* (gbest - X);
        X = X + V;

        % Apply search space constraints
        X = max(min(X, bounds(2)), bounds(1));

        % Evaluate objective function
        fitness = ackley(X);

        % Update best individual positions
        update_mask = fitness < pbest_fitness;
        pbest(update_mask, :) = X(update_mask, :);
        pbest_fitness(update_mask) = fitness(update_mask);

        % Update global best
        [new_gbest_fitness, idx] = min(pbest_fitness);
        if new_gbest_fitness < gbest_fitness
            gbest_fitness = new_gbest_fitness;
            gbest = pbest(idx, :);
        end

        % Display progress
        fprintf('Iteration %d: Best value found = %.5f\n', iter, gbest_fitness);
    end
```

```
    fprintf('Best solution found: [%.5f, %.5f]\n', gbest(1), gbest(2));
end

function f = ackley(X)
    a = 20;
    b = 0.2;
    c = 2 * pi;

    n = size(X, 2);
    sum1 = sum(X.^2, 2);
    sum2 = sum(cos(c * X), 2);

    f = -a * exp(-b * sqrt(sum1 / n)) - exp(sum2 / n) + a + exp(1);
end
```

In the same way, the PSO-CHAOS algorithm was implemented to complete the 3 initialization modifications proposed for PSO.

```
function pso_ackley_chaos()
    % PSO Parameters
    num_particles = 30; % Number of particles
    num_dimensions = 2; % Problem dimensions
    max_iterations = 100; % Maximum iterations
    w = 0.7;    % Inertia weight
    c1 = 1.5;  % Cognitive coefficient
    c2 = 1.5;  % Social coefficient
    bounds = [-32.768, 32.768]; % Ackley function bounds

    % Initialization with Chaos-Based Sampling (Logistic Map)
    mu = 3.99; % Chaos parameter (near chaotic regime)
    z = rand(num_particles, 1); % Initial random values
    X = zeros(num_particles, num_dimensions);

    for i = 1:num_dimensions
        z = mu * z .* (1 - z); % Logistic map iteration
        X(:, i) = bounds(1) + (bounds(2) - bounds(1)) * z;
    end

    V = zeros(num_particles, num_dimensions); % Initial velocities

    % Initial evaluation
    fitness = ackley(X);
    pbest = X; % Best individual positions
    pbest_fitness = fitness;

    % Global best
    [gbest_fitness, idx] = min(pbest_fitness);
    gbest = pbest(idx, :);

    % PSO iterations
    for iter = 1:max_iterations
        % Update velocity and position
        r1 = rand(num_particles, num_dimensions);
        r2 = rand(num_particles, num_dimensions);
        V = w * V + c1 * r1 .* (pbest - X) + c2 * r2 .* (gbest - X);
        X = X + V;

        % Apply search space constraints
        X = max(min(X, bounds(2)), bounds(1));

        % Evaluate objective function
        fitness = ackley(X);

        % Update best individual positions
```

```
        update_mask = fitness < pbest_fitness;
        pbest(update_mask, :) = X(update_mask, :);
        pbest_fitness(update_mask) = fitness(update_mask);

        % Update global best
        [new_gbest_fitness, idx] = min(pbest_fitness);
        if new_gbest_fitness < gbest_fitness
            gbest_fitness = new_gbest_fitness;
            gbest = pbest(idx, :);
        end

        % Display progress
        fprintf('Iteration %d: Best value found = %.5f\n', iter, gbest_fitness);
    end

    fprintf('Best solution found: [%.5f, %.5f]\n', gbest(1), gbest(2));
end

function f = ackley(X)
    a = 20;
    b = 0.2;
    c = 2 * pi;

    n = size(X, 2);
    sum1 = sum(X.^2, 2);
    sum2 = sum(cos(c * X), 2);

    f = -a * exp(-b * sqrt(sum1 / n)) - exp(sum2 / n) + a + exp(1);
end
```

The original DE algorithm was also modified with the 3 proposed methods to empirically analyze the initialization effect with different methods, resulting in the following code implementations.

The following code corresponds to DE-LHS.

```
function de_ackley_lhs()
    % Differential Evolution (DE) Parameters
    pop_size = 30; % Population size
    num_dimensions = 2; % Problem dimensions
    max_generations = 100; % Maximum generations
    F = 0.5; % Mutation factor
    CR = 0.9; % Crossover probability
    bounds = [-32.768, 32.768]; % Ackley function bounds

    % Initialization using Latin Hypercube Sampling (LHS)
    pop = lhsdesign(pop_size, num_dimensions) * (bounds(2) - bounds(1)) + bounds
    (1);
    fitness = ackley(pop);

    % Evolution process
    for gen = 1:max_generations
        for i = 1:pop_size
            % Mutation: Select three distinct individuals
            idxs = randperm(pop_size, 3);
            while any(idxs == i)
                idxs = randperm(pop_size, 3);
            end
            donor = pop(idxs(1), :) + F * (pop(idxs(2), :) - pop(idxs(3), :));
            donor = max(min(donor, bounds(2)), bounds(1)); % Enforce bounds

            % Crossover
```

```
            j_rand = randi(num_dimensions);
            trial = pop(i, :);
            for j = 1:num_dimensions
                if rand < CR || j == j_rand
                    trial(j) = donor(j);
                end
            end

            % Selection
            trial_fitness = ackley(trial);
            if trial_fitness < fitness(i)
                pop(i, :) = trial;
                fitness(i) = trial_fitness;
            end
        end

        % Display progress
        fprintf('Generation %d: Best fitness = %.5f\n', gen, min(fitness));
    end

    % Best solution found
    [best_fitness, best_idx] = min(fitness);
    best_solution = pop(best_idx, :);
    fprintf('Best solution found: [%.5f, %.5f] with fitness %.5f\n',
     best_solution(1), best_solution(2), best_fitness);
end

function f = ackley(X)
    a = 20;
    b = 0.2;
    c = 2 * pi;
    n = size(X, 2);
    sum1 = sum(X.^2, 2);
    sum2 = sum(cos(c * X), 2);
    f = -a * exp(-b * sqrt(sum1 / n)) - exp(sum2 / n) + a + exp(1);
end
```

The code below continues with the modification process, resulting in the DE-OBL algorithm.

```
function de_ackley_obl()
    % Differential Evolution (DE) Parameters
    pop_size = 30; % Population size
    num_dimensions = 2; % Problem dimensions
    max_generations = 100; % Maximum generations
    F = 0.5; % Mutation factor
    CR = 0.9; % Crossover probability
    bounds = [-32.768, 32.768]; % Ackley function bounds

    % Initialization using Opposition-Based Learning (OBL)
    pop = rand(pop_size, num_dimensions) * (bounds(2) - bounds(1)) + bounds(1);
    pop_opposite = bounds(1) + bounds(2) - pop;

    % Evaluate fitness for both original and opposite population
    fitness = ackley(pop);
    fitness_opposite = ackley(pop_opposite);

    % Select the best individuals
    mask = fitness_opposite < fitness;
    pop(mask, :) = pop_opposite(mask, :);
    fitness(mask) = fitness_opposite(mask);

    % Evolution process
    for gen = 1:max_generations
        for i = 1:pop_size
```

```
            % Mutation: Select three distinct individuals
            idxs = randperm(pop_size, 3);
            while any(idxs == i)
                idxs = randperm(pop_size, 3);
            end
            donor = pop(idxs(1), :) + F * (pop(idxs(2), :) - pop(idxs(3), :));
            donor = max(min(donor, bounds(2)), bounds(1)); % Enforce bounds

            % Crossover
            j_rand = randi(num_dimensions);
            trial = pop(i, :);
            for j = 1:num_dimensions
                if rand < CR || j == j_rand
                    trial(j) = donor(j);
                end
            end

            % Selection
            trial_fitness = ackley(trial);
            if trial_fitness < fitness(i)
                pop(i, :) = trial;
                fitness(i) = trial_fitness;
            end
        end

        % Display progress
        fprintf('Generation %d: Best fitness = %.5f\n', gen, min(fitness));
    end

    % Best solution found
    [best_fitness, best_idx] = min(fitness);
    best_solution = pop(best_idx, :);
    fprintf('Best solution found: [%.5f, %.5f] with fitness %.5f\n',
     best_solution(1), best_solution(2), best_fitness);
end

function f = ackley(X)
    a = 20;
    b = 0.2;
    c = 2 * pi;
    n = size(X, 2);
    sum1 = sum(X.^2, 2);
    sum2 = sum(cos(c * X), 2);
    f = -a * exp(-b * sqrt(sum1 / n)) - exp(sum2 / n) + a + exp(1);
end
```

Finally, the last variation of the DE algorithm was the DE-CHAOS combination.

```
function de_ackley_chaos()
    % Differential Evolution (DE) Parameters
    pop_size = 30; % Population size
    num_dimensions = 2; % Problem dimensions
    max_generations = 100; % Maximum generations
    F = 0.5; % Mutation factor
    CR = 0.9; % Crossover probability
    bounds = [-32.768, 32.768]; % Ackley function bounds

    % Initialization using Chaos-Based Sampling (Logistic Map)
    mu = 3.99;
    z = rand(pop_size, 1);
    pop = zeros(pop_size, num_dimensions);
    for i = 1:num_dimensions
        z = mu * z .* (1 - z);
        pop(:, i) = bounds(1) + (bounds(2) - bounds(1)) * z;
    end
```

```
    fitness = ackley(pop);

    % Evolution process
    for gen = 1:max_generations
        for i = 1:pop_size
            % Mutation: Select three distinct individuals
            idxs = randperm(pop_size, 3);
            while any(idxs == i)
                idxs = randperm(pop_size, 3);
            end
            donor = pop(idxs(1), :) + F * (pop(idxs(2), :) - pop(idxs(3), :));
            donor = max(min(donor, bounds(2)), bounds(1)); % Enforce bounds

            % Crossover
            j_rand = randi(num_dimensions);
            trial = pop(i, :);
            for j = 1:num_dimensions
                if rand < CR || j == j_rand
                    trial(j) = donor(j);
                end
            end

            % Selection
            trial_fitness = ackley(trial);
            if trial_fitness < fitness(i)
                pop(i, :) = trial;
                fitness(i) = trial_fitness;
            end
        end

        % Display progress
        fprintf('Generation %d: Best fitness = %.5f\n', gen, min(fitness));
    end

    % Best solution found
    [best_fitness, best_idx] = min(fitness);
    best_solution = pop(best_idx, :);
    fprintf('Best solution found: [%.5f, %.5f] with fitness %.5f\n',
     best_solution(1), best_solution(2), best_fitness);
end

function f = ackley(X)
    a = 20;
    b = 0.2;
    c = 2 * pi;
    n = size(X, 2);
    sum1 = sum(X.^2, 2);
    sum2 = sum(cos(c * X), 2);
    f = -a * exp(-b * sqrt(sum1 / n)) - exp(sum2 / n) + a + exp(1);
end
```

All algorithms share common initialization parameters (e.g., number of solutions, dimensions, search space bounds, stopping criterion). To compare their performance, two experimental tests were conducted—evaluating three PSO versions with LHS, OBL, and CHAOS and separately testing DE with the same initializations.

5.4 Test Setup and Configuration

This section outlines the experimental setup for Particle Swarm Optimization (PSO) and Differential Evolution (DE). To ensure consistency across different initialization methods, both algorithms were tested on a 30-dimensional Ackley function with defined search boundaries.

- Population size: 50.
- Runs per algorithm: 30 (for statistical significance).
- Initialization strategies: Various methods assessed for impact on performance.

The following tables summarize the parameter settings for PSO and DE in this study.

Table 5.1: Initialization parameter configuration for PSO.

Parameter	Value
Number of particles	50
Problem dimensions	30
Maximum iterations	1000
Number of independent runs	30
Search boundaries	[–30, 30]
Inertia weight (w)	0.7
Cognitive coefficient (c1)	1.5
Social coefficient (c2)	1.5
Initialization methods	Latin Hypercube, Opposition-Based, Chaos-Based

Table 5.2: Initialization parameter configuration for Differential Evolution (DE).

Parameter	Value
Population size	50
Problem dimensions	30
Maximum iterations	1000
Mutation factor (F)	0.5
Crossover rate (CR)	0.7
Search boundaries	[-30, 30]
Number of independent runs	30
Initialization methods	Latin Hypercube, Opposition-Based Learning, Chaos-Based Sampling

5.5 Experimental Results

Extensive experiments with PSO and DE on the Ackley function assessed the effectiveness of different initialization methods. Results highlight their impact on algorithm performance, using statistical measures like mean best-found fitness and standard deviation across multiple runs. Additionally, convergence trends provide insights into stability and efficiency.

5.6 Discussion

This study evaluates LHS, OBL, and CHAOS initialization strategies in PSO and DE optimization. Using the Ackley function in a 30-dimensional search space ([–30, 30]), each algorithm was run 30 times independently to ensure statistical validity.

PSO was configured with inertia weight (0.7), cognitive/social coefficients (1.5), swarm size (50), and 1000 iterations. DE used a mutation factor (0.5), crossover rate (0.7), and population size (50). Experimental results provided insights into initialization technique effectiveness.

Table 5.3 shows PSO performance across three initialization strategies. PSO-OBL achieved the best results, with the lowest mean fitness and minimal variance. PSO-LHS performed slightly worse, while PSO-CHAOS showed significantly poorer optimization, indicating its ineffectiveness in this test case.

Table 5.4 highlights DE performance across three initialization strategies. DE-OBL achieved near-zero optimal fitness with no deviation, outperforming the others. DE-LHS performed similarly, though with a slightly higher mean error. DE-CHAOS, while effective, showed the highest variance, suggesting OBL offers a more stable and reliable starting point for superior convergence.

Figures 5.6 and 5.7 highlight algorithm convergence trends. PSO-OBL and DE-OBL consistently achieved faster and better solutions than other methods. DE-OBL, in particular, demonstrated rapid fitness reduction, confirming its robustness as an initialization strategy.

Table 5.3: Performance comparison on PSO variants (Best found fitness).

Method	Mean	Std Dev
PSO-LHS	2.56652	2.42622
PSO-OBL	**2.21325**	**1.30934**
PSO-CHAOS	13.96326	7.65282

Table 5.4: Performance comparison on DE variants (Best found fitness).

Method	Mean	Std Dev
DE-LHS	1.0×10^{-5}	0.0
DE-OBL	0.0	0.0
DE-CHAOS	2.0×10^{-5}	3.0×10^{-5}

Overall, our findings indicate that the choice of initialization method plays a crucial role in the optimization performance of both PSO and DE. Among the evaluated strategies, Opposition-Based Learning (OBL) consistently delivered the most accurate

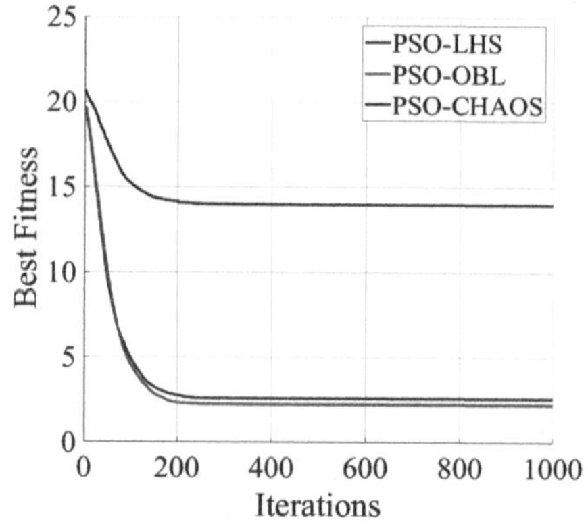

Figure 5.6: Convergence curves for PSO on Ackley function.

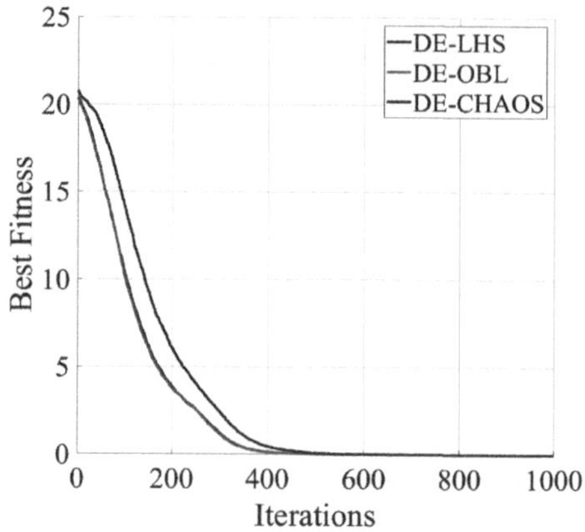

Figure 5.7: Convergence curves for DE on Ackley function.

and stable results for both algorithms. While Latin Hypercube Sampling (LHS) performed well, it proved slightly less effective. In contrast, Chaos-Based Sampling (CHAOS) exhibited greater variability and lower performance, particularly for PSO. These results underscore the importance of choosing an initialization technique that aligns with the specific characteristics of the optimization algorithm.

References

[1] M. Črepinšek, S.-H. Liu, and M. Mernik, "Exploration and exploitation in evolutionary algorithms: A survey," *ACM Computing Surveys (CSUR)*, vol. 45, no. 3, pp. 1–33, 2013.

[2] M. Srinivas and L. M. Patnaik, "Genetic algorithms: A survey," *Computer*, vol. 27, no. 6, pp. 17–26, 1994.

[3] R. Storn and K. Price, "Differential evolution–a simple and efficient heuristic for global optimization over continuous spaces," *Journal of Global Optimization*, vol. 11, no. 4, pp. 341–359, 1997.

[4] M. Stein, "Large sample properties of simulations using latin hypercube sampling," *Technometrics*, vol. 29, no. 2, pp. 143–151, 1987.

[5] H. R. Tizhoosh, "Opposition-based learning: A new scheme for machine intelligence," in *International Conference on Computational Intelligence for Modelling, Control and Automation and International Conference on Intelligent Agents, Web Technologies and Internet Commerce (CIMCA-IAWTIC'06)*, vol. 1. IEEE, 2005, pp. 695–701.

[6] Y. Zhang, J. Lu, C. Zhao, Z. Li, and J. Yan, "Chaos optimization algorithms: A survey," *International Journal of Bifurcation and Chaos*, vol. 34, no. 16, p. 2450205, 2024.

[7] A. Mashuri, N. H. Adenan, N. S. Abd Karim, S. W. Tho, and Z. Zeng, "Application of chaos theory in different fields-a literature review," *Journal of Science and Mathematics Letters*, vol. 12, no. 1, pp. 92–101, 2024.

[8] C. Oestreicher, "A history of chaos theory," *Dialogues in Clinical Neuroscience*, vol. 9, no. 3, pp. 279–289, 2007.

Chapter 6

The Importance of Diversity in Metaheuristics

6.1 Introduction

In recent years, metaheuristic search algorithms have gained popularity for solving a wide range of optimization problems across diverse domains. These problems span engineering design [1, 2], where optimization enhances system performance and efficiency; digital image processing [3, 4], aiding in tasks like segmentation and enhancement; and computer vision [5, 6], contributing to object recognition and tracking. Additionally, metaheuristics play a crucial role in networks and communications [7, 8], optimizing routing and resource allocation; power and energy management, improving distribution and consumption efficiency; and data analysis and machine learning [9, 10], enhancing model training and feature selection. These algorithms are also widely applied in robotics for path planning and control [11, 12], and in medical diagnosis [13, 14], assisting in interpreting complex diagnostic data and improving decision-making processes. The versatility and effectiveness of metaheuristic algorithms make them indispensable tools for tackling complex optimization challenges across various fields [15].

Most metaheuristic methods rely on population-based search schemes, where a population of search agents (or individuals) applies specific rules to explore candidate solutions within a feasible solution space. These optimization frameworks offer several advantages. A key benefit is the interaction among individuals, which facilitates knowledge exchange between solutions. Additionally, maintaining population

DOI: 10.1201/9781032723563-6

diversity is essential for efficiently exploring the search space and overcoming local optima [16].

Metaheuristic search methods vary widely in design and application, prompting the question: What most influences their effectiveness? A widely accepted principle is that balancing exploration and exploitation improves performance [17]. However, the specifics of this balance remain unclear [18, 19], and categorizing search operators and strategies is often ambiguous, as they may serve both exploration and exploitation [20].

Due to the uncertainty in defining exploration-exploitation balance, various metrics have been proposed to quantify these aspects by monitoring population diversity [18, 19, 20, 21, 22, 23, 24, 25]. Despite numerous indices and ongoing developments, no definitive method objectively measures exploration and exploitation rates in metaheuristic algorithms.

A key metric in metaheuristics, dimension-wise diversity [24, 26] quantifies exploration by calculating the average sum of distances between solutions and the median of each dimension. Exploration percentage is obtained by dividing diversity by its maximum observed value, with exploitation as its inverse. This index helps assess search agent distribution, indicating whether an algorithm focuses more on exploration or exploitation. Despite its usefulness, it remains underutilized in metaheuristic research.

Leveraging diversity is key to optimizing performance in metaheuristics and solving complex problems efficiently. Ongoing research on exploration-exploitation balance metrics continues to refine these powerful optimization methods.

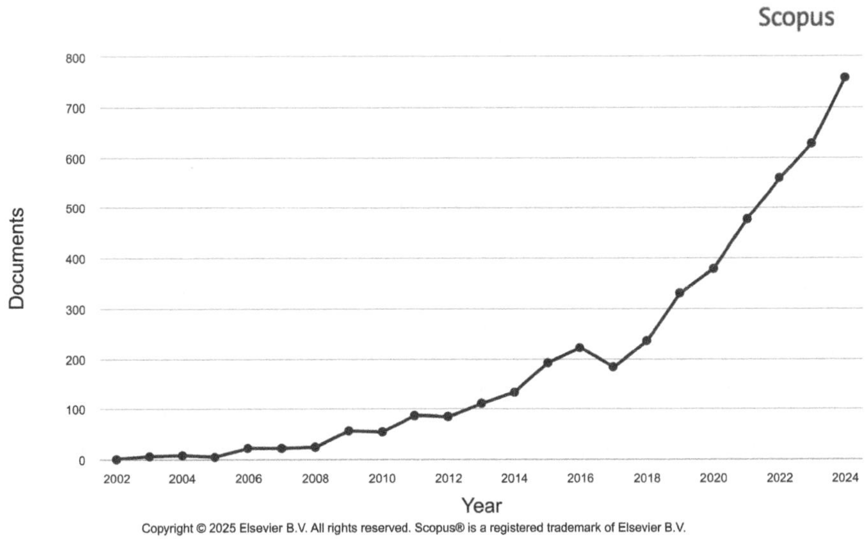

Figure 6.1: Documents by year related to diversity approaches.

This chapter explores exploration and exploitation in metaheuristic algorithms, followed by an analysis of the most common diversity metric and its role in balancing search efficiency. Lastly, key applications in optimization demonstrate how diversity measures enhance performance and robustness across various domains, providing insights into achieving optimal solutions.

6.2 Exploration and Exploitation in Metaheuristics

In metaheuristic algorithms, exploration ensures diverse search across the space, while exploitation refines solutions in promising regions. Research highlights their impact on convergence rate—exploitation accelerates optimization but risks local optima, whereas exploration enhances global search but slows convergence [27].

The mechanisms driving exploration and exploitation in metaheuristic optimization remain a subject of debate [28]. While balancing both is widely recognized as key to optimal performance, defining the ideal ratio for efficiency is challenging, given the diversity of search methods. Understanding how these strategies influence the search process is crucial for designing effective optimization approaches.

In population-based search methods, selection mechanisms maintain a balance between elitism and diversity [29]. Greedy selection ensures top solutions persist across generations, accelerating convergence. In Differential Evolution (DE), individual selection accepts only improved solutions [30], refining the exploration-exploitation balance by guiding diverse populations toward better outcomes [31].

Some search algorithms lack selection strategies, accepting all new solutions regardless of quality. To balance exploration and exploitation, PSO [32] and GWO [33] use attraction operators, guiding solutions toward promising individuals. In PSO, particles follow both their personal best and the global best, ensuring a balanced attraction mechanism but requiring additional memory.

Some algorithms utilize complex attraction strategies, considering multiple attractors within a subset or the entire population. In Firefly Algorithm (FA) [34], attractiveness depends on solution quality and distance, with farther individuals exerting lower attraction. Gravitational Search Algorithm (GSA) [35] relies on gravitational force, where fitness values and distances dictate attraction strength.

Some search methods rely on random walks or distance-based criteria rather than attraction mechanisms. In Genetic Algorithms (GA), solutions emerge through crossover (mixing information) and mutation (adding perturbations). While crossover initially aids exploration, its effect diminishes as population diversity decreases, shifting toward exploitation. Conversely, large mutation perturbations enhance exploration. Since crossover and mutation roles shift, their classification depends on rate adjustments within the search process.

Some metaheuristic algorithms dynamically adjust exploration-exploitation rates based on iteration progress and stop criteria. This parameter tuning helps prevent premature convergence, but frequent adjustments may negatively affect convergence speed, making them unsuitable in certain cases [20].

Balancing exploration and exploitation in metaheuristic algorithms relies on effective diversity measurements. These metrics assess solution spread, identifying over-exploitation (too localized search) or over-exploration (excessive dispersion). By quantifying diversity, researchers fine-tune search parameters for optimal performance. For instance, dimension-wise diversity calculates the average solution distance, offering insights into population distribution. Tracking diversity over time allows adaptive search adjustments, ensuring efficient optimization across complex landscapes. Thus, robust diversity measures are essential for managing exploration-exploitation dynamics in metaheuristic frameworks.

6.3 Applications in the Optimization Field

Balancing exploration and exploitation is pivotal to metaheuristic success. The dimension-wise diversity metric serves as a structured tool for evaluating search dynamics. By measuring solution dispersion across dimensions, it offers insights into an algorithm's efficiency in exploring the search space while optimizing promising regions.

Dimension-wise diversity plays a vital role in engineering design, image processing, machine learning, and other domains. This section examines its impact on metaheuristic algorithm performance, exploring case studies and theoretical analyses to highlight its role in optimizing complex systems and solving diverse challenges.

Understanding dimension-wise diversity allows researchers to refine metaheuristic algorithms, efficiently navigate search spaces, and achieve superior optimization solutions. This section explores practical examples and methodologies, showcasing its versatility and effectiveness in advancing optimization.

6.3.1 *Measuring Balance in Successful Algorithms*

Metaheuristic algorithms rely on exploration to scan diverse search regions and exploitation to refine promising solutions. Striking an optimal balance is crucial for efficient convergence and finding high-quality solutions in optimization problems.

This subsection analyzes the exploration-exploitation balance across popular metaheuristic algorithms, including ABC [36, 37], BA [38], CMA-ES [39], CSA [40], DE [30], FA [34], GWO [33], SSO [41], TLBO [42], MFO [43], PSO [44], and WOA [45]. These algorithms showcase diverse search mechanisms and design methodologies, offering a comprehensive perspective on balance dynamics in metaheuristic optimization.

Each algorithm is tested on 43 benchmark functions covering unimodal, multimodal, composite, and shifted functions to assess performance. The exploration-exploitation balance is quantified using dimension-wise diversity and other evaluation metrics.

Exploration-Exploitation Analysis in Metaheuristics:

This study systematically measures, analyzes, and evaluates exploration-exploitation dynamics across metaheuristic algorithms. It examines balance trade-offs, assesses performance effectiveness, and identifies common patterns for algorithm improvement. The findings provide guidance for fine-tuning optimization methods, enhancing their applicability across diverse fields.

6.3.1.1 Comparative Analysis

Five indicators assess algorithm performance: average (AB), median (MD), standard deviation (SD) of best solutions from 30 runs, exploration percentage (%XPL), and exploitation percentage (%XPT). The Friedman test [46], a non-parametric statistical test, ranks algorithms by benchmark performance. A p-value below 0.01 or 0.05 indicates significant differences, rejecting the null hypothesis [46].

Table 6.1 summarizes algorithm traits influencing exploration, exploitation, or both. It details selection mechanisms, attraction operators, and whether iteration count is factored into the search strategy.

Table 6.1: Observable characteristics of selected algorithms.

Algorithms	Selection Mechanism	Attraction Operators	Iteration Dependent
ABC	Ind. Greedy	Multiple	NO
BA	Ind. Greedy	Global Best	NO
CMA-ES	Non-Greedy	N/A	NO
CSA	Non-Greedy	Personal	NO
DE	Ind. Greedy	N/A	NO
FA	Non-Greedy	Multiple	NO
GWO	Non-Greedy	Multiple	NO
MFO	Greedy	Multiple	YES
PSO	Non-Greedy	Multiple	NO
SSO	Non-Greedy	Multiple	NO
TLBO	Greedy	Multiple	NO
WOA	Non-Greedy	Global Best	NO

The chosen algorithms are notable for performance, novelty, or accessible source code. Table 6.2 outlines their settings sourced from literature. Due to the extensive graphical data on exploration-exploitation rates across benchmarks, results are organized by function types, each subsection analyzing algorithm behaviors and effectiveness.

Table 6.2: Parameter settings for each algorithm.

Algorithm	Parameters
ABC	Colony population = 124, Onlooker bees = 62, Employed bees = 62, Scout bee = 1, limit = 100 [36].
BA	Bat population = 50, alpha = 0.9, gamma = 0.9 [38].
CMA-ES	The algorithm has been configured according to the guidelines provided by its author [47].
CSA	Crow population = 50, Flight length = 2, awareness probability = 0.1 [40].
DE	The weight parameter is set to F=0.75 while the crossover probability is configured to CR = 0.2 [48].
FA	The parameters set up for the randomness factor and the light absorption coefficient are set to $\alpha = 0.98^{it}$ and $\gamma = 1.0$ respectively, where it is the iteration number [49].
GWO	Wolf population = 30 [33].
MFO	The number of flames is set as $N_{flames} = round\left(\frac{N_{pop} - k \times (N_{pop} - 1)}{k_{max}}\right)$ where N_{pop} denotes the population size, k the current iteration and k_{max} the maximum number of iterations [43].
PSO	The learning factors are set to $c1 = 2$ and $c2 = 2$. On the other hand, the inertia weight factor is configured to decrease linearly from 0.9 to 0.2 as the process evolves [50].
SSO	Number of females $NFem = 50 \times rand$, Number of males $NMa = 50 - NFem$ where $rand$ is a number between 0.65 and 0.9 [51].
TLBO	Population = 50 [42].
WOA	Whale population = 30 [45].

Multimodal Functions: Analyzes the balance of metaheuristic methods when applied to multimodal functions. Unimodal Functions: Evaluates the algorithms' performance on unimodal functions. Hybrid Functions: Discusses balance responses over hybrid functions. Shifted Functions: Analyzes results over shifted functions. Random Search Algorithm: Examines the balance response of the Random Search algorithm. Diversity Analysis: Conducts a detailed diversity analysis. These sections provide insights into the balance between exploration and exploitation, demonstrating the strengths and weaknesses of each algorithm under various conditions. This understanding helps researchers and practitioners select and fine-tune metaheuristic algorithms for specific optimization challenges.

6.3.1.2 Multimodal Functions

This subsection focuses on the results for multimodal functions, characterized by their multiple local optima. Table 6.3 compares the performance of the selected algorithms on these functions, while Table 6.4 presents the statistical analysis results using the

Friedman test. The best results are highlighted in bold. The analysis suggests that the best performing algorithms are CMAES, TLBO, and WOA.

Table 6.3: Best solutions found and balance employed in each multimodal function.

		ABC	BA	CMAES	CSA	DE	FA	GWO	MFO	PSO	SSO	TLBO	WOA
	AB	1.72E+01	1.25E+01	2.06E+00	1.71E+01	5.12E-03	1.77E+00	1.01E-14	9.22E+00	5.44E+00	3.54E-01	5.33E-15	**3.08E-15**
F1	**MD**	1.73E+01	1.25E+01	7.11E-15	1.72E+01	4.84E-03	1.81E+00	8.88E-15	1.39E+01	2.38E-02	3.48E-01	5.33E-15	**3.55E-15**
	SD	4.69E-01	1.07E+00	6.30E+00	4.15E-01	9.34E-04	1.43E-01	3.24E-15	8.37E+00	7.98E+00	5.43E-02	**1.81E-15**	2.59E-15
	%XPL	87.497	0.9934	2.0427	98.0255	13.4553	14.4189	0.4716	6.7257	44.4648	6.9182	1.0521	1.8393
	%XPT	12.503	99.0066	97.9573	1.9745	86.5447	85.5811	99.5284	93.2743	55.5352	93.0818	98.9479	98.1607
	AB	2.86E+05	2.52E+01	6.67E-01	1.69E+05	1.46E+00	4.17E+00	**6.67E-01**	4.28E+04	1.80E+04	3.69E+00	**6.67E-01**	**6.67E-01**
F2	**MD**	2.74E+05	3.92E+00	6.67E-01	1.70E+05	1.38E+00	3.92E+00	**6.67E-01**	8.24E+01	3.21E+02	3.34E+00	**6.67E-01**	**6.67E-01**
	SD	7.76E+04	4.93E+01	6.03E-15	4.70E+04	3.65E-01	1.20E+00	4.94E-06	1.16E+05	4.27E+04	1.46E+00	**3.41E-14**	8.19E-05
	%XPL	93.9632	3.1009	2.4524	98.5658	15.3386	19.1611	0.515	5.7613	45.8991	19.1499	1.0151	2.5674
	%XPT	6.0368	96.8991	97.5476	1.4342	84.6614	80.8389	99.485	94.2387	54.1009	80.8501	98.9849	97.4326
	AB	1.37E+02	7.76E+01	1.31E-03	1.70E+02	6.01E-03	1.10E+00	1.05E-03	1.19E-02	9.03E+00	1.66E-02	**0.00E+00**	1.34E-03
F3	**MD**	1.39E+02	6.65E+01	0.00E+00	1.73E+02	3.98E-03	1.10E+00	0.00E+00	9.87E-03	8.86E-03	1.40E-02	**0.00E+00**	0.00E+00
	SD	1.61E+01	2.51E+01	3.47E-03	2.03E+01	5.66E-03	1.24E-02	4.40E-03	1.32E-02	2.75E+01	1.05E-02	**0.00E+00**	7.35E-03
	%XPL	78.6281	2.94	2.0703	98.3754	9.9041	16.8046	0.464	4.9714	38.749	6.6239	0.9905	1.7932
	%XPT	21.3719	97.06	97.9297	1.6246	90.0959	83.1954	99.536	95.0286	61.251	93.3761	99.0095	98.2068
	AB	4.98E-01	3.21E-06	5.29E-111	1.18E-01	2.42E-16	1.05E-09	4.92E-217	4.53E-08	2.17E-10	4.68E-05	1.98E-236	**0.00E+00**
F4	**MD**	4.96E-01	3.07E-06	1.04E-112	1.07E-01	2.12E-16	8.43E-10	1.26E-221	1.19E-09	7.22E-12	3.98E-05	2.49E-238	**0.00E+00**
	SD	1.56E-01	1.00E-06	1.96E-110	4.14E-02	1.68E-16	7.67E-10	0.00E+00	1.23E-07	9.22E-10	3.14E-05	0.00E+00	**0.00E+00**
	%XPL	96.9063	7.1664	2.4101	98.1739	15.5385	18.7493	0.5353	7.6471	49.5202	55.0966	0.9975	2.3214
	%XPT	3.0937	92.8336	97.5899	1.8261	84.4615	81.2507	99.4647	92.3529	50.4798	44.9034	99.0025	97.6786
	AB	8.36E+01	7.82E+00	2.12E-01	5.83E+01	**2.52E-05**	3.47E-01	1.11E+00	2.48E+01	4.94E+00	3.09E-01	4.74E-01	9.66E-02
F5	**MD**	8.47E+01	6.57E+00	1.50E-32	5.84E+01	**2.43E-05**	9.07E-02	1.09E+00	2.33E+01	3.50E+00	1.27E-01	4.92E-01	9.51E-02
	SD	1.21E+01	4.65E+00	5.06E-01	8.60E+00	**8.90E-06**	5.24E-01	2.14E-01	1.01E+01	4.94E+00	4.17E-01	2.22E-01	1.02E-01
	%XPL	95.0028	2.59	2.6241	98.1753	17.332	17.5826	2.6767	5.0162	47.1323	19.6133	1.2255	6.3444
	%XPT	4.9972	97.41	97.3759	1.8247	82.668	82.4174	97.3233	94.9838	52.8677	80.3867	98.7745	93.6556
	AB	1.91E-01	1.81E-04	5.57E-27	9.36E-02	**0.00E+00**	9.52E-08	7.04E-12	7.12E-07	0.00E+00	9.32E-07	**0.00E+00**	**0.00E+00**
F6	**MD**	1.88E-01	5.85E-15	0.00E+00	9.09E-02	**0.00E+00**	8.42E-08	3.69E-12	0.00E+00	0.00E+00	7.50E-07	**0.00E+00**	**0.00E+00**
	SD	4.67E-02	9.76E-04	3.05E-26	2.18E-02	**0.00E+00**	4.01E-08	7.35E-12	3.90E-06	0.00E+00	6.86E-07	**0.00E+00**	**0.00E+00**
	%XPL	95.9426	3.501	2.9087	98.4887	70.1519	13.0572	25.8766	4.0018	0.5549	20.5388	14.9194	24.225
	%XPT	4.0574	96.499	97.0913	1.5113	29.8481	86.9428	74.1234	95.9982	99.4451	79.4612	85.0806	75.775
	AB	3.79E+04	6.36E+04	1.63E-87	5.81E+01	1.69E-111	1.03E-38	**0.00E+00**	4.66E-206	7.25E-33	4.39E-17	**0.00E+00**	**0.00E+00**
F7	**MD**	5.61E+03	1.41E-04	1.80E-121	1.14E+01	1.17E-116	1.69E-42	**0.00E+00**	8.15E-225	1.36E-90	3.59E-22	**0.00E+00**	**0.00E+00**
	SD	1.10E+05	3.21E+05	6.12E-87	1.24E+02	8.88E-111	5.60E-38	**0.00E+00**	0.00E+00	3.97E-32	1.46E-16	**0.00E+00**	**0.00E+00**
	%XPL	86.4106	1.0759	3.5939	98.0479	12.4436	22.948	0.6234	6.7044	37.5761	23.0048	1.3033	1.6233
	%XPT	13.5894	98.9241	96.4061	1.9521	87.5564	77.052	99.3766	93.2956	62.4239	76.9952	98.6967	98.3767
	AB	1.15E+08	9.53E+06	**7.69E+01**	4.59E+07	2.00E+05	1.45E+03	2.27E+06	8.66E+01	9.20E+02	8.27E+01	2.64E+02	4.52E+02
F8	**MD**	1.16E+08	1.03E+07	**7.65E+01**	4.51E+07	1.97E+05	1.22E+03	2.00E+06	8.73E+01	3.23E+02	8.20E+01	7.82E+01	1.45E+02
	SD	3.13E+07	3.34E+06	**3.07E+00**	9.43E+06	6.77E+04	9.15E+02	1.44E+06	3.74E+00	1.60E+03	2.51E+00	7.47E+02	1.19E+03
	%XPL	95.5115	3.4265	2.4201	98.5632	39.5521	18.013	7.3906	6.4148	42.9836	8.8978	2.712	10.5891
	%XPT	4.4885	96.5735	97.5799	1.4368	60.4479	81.987	92.6094	93.5852	57.0164	91.1022	97.288	89.4109
	AB	2.09E+08	2.82E+06	1.04E+02	7.46E+07	1.92E+04	1.22E+03	1.40E+05	**9.14E+01**	7.75E+02	1.21E+02	1.36E+02	4.60E+03
F9	**MD**	2.08E+08	2.25E+06	1.04E+02	7.82E+07	1.75E+04	9.04E+02	1.25E+05	**9.18E+01**	2.38E+02	1.20E+02	1.22E+02	1.64E+02
	SD	6.05E+07	2.48E+06	4.61E+00	1.87E+07	7.84E+03	1.11E+03	7.09E+04	**3.72E+00**	1.54E+03	5.36E+00	5.87E+01	1.95E+04
	%XPL	95.3091	3.5499	2.3544	99.0277	24.8489	19.8155	3.7775	6.7386	48.9883	9.1842	2.0446	6.8941
	%XPT	4.6909	96.4501	97.6456	0.9723	75.1511	80.1845	96.2225	93.2614	51.0117	90.8158	97.9554	93.1059
	AB	6.18E+84	9.59E+83	1.89E+83	5.10E+83	3.06E+81	8.60E+81	3.60E+81	**1.66E+81**	4.40E+81	4.26E+81	1.82E+82	1.29E+82
F10	**MD**	3.86E+84	8.54E+83	1.02E+83	4.32E+83	1.25E+81	6.39E+81	7.87E+80	**2.36E+80**	2.47E+81	9.35E+80	1.07E+82	3.02E+81
	SD	6.29E+84	7.77E+83	2.81E+83	3.87E+83	6.49E+81	7.58E+81	6.65E+81	**3.25E+81**	5.29E+81	8.50E+81	1.84E+82	2.56E+82
	%XPL	97.9017	1.0069	29.1786	99.1435	89.9032	51.7721	14.797	8.4823	60.202	33.6651	84.3472	12.5809
	%XPT	2.0983	98.9931	70.8214	0.8565	10.0968	48.2279	85.203	91.5177	39.798	66.3349	15.6528	87.4191
	AB	1.55E+83	2.85E+63	2.04E+02	2.90E+75	1.25E+02	8.65E+28	2.71E+02	7.61E+37	2.15E+49	8.15E+23	1.64E+02	**1.41E+02**
F11	**MD**	8.93E+81	1.60E+54	3.03E+01	7.22E+73	9.93E+01	1.73E+21	1.15E+02	2.09E+25	4.00E+32	5.96E+17	5.41E+01	**5.09E+01**
	SD	4.24E+83	1.56E+64	4.10E+02	5.96E+75	7.70E+01	4.28E+29	3.81E+02	4.17E+38	1.18E+50	2.99E+24	2.53E+02	**2.52E+02**
	%XPL	97.5495	2.4453	0.5018	98.6073	20.5013	23.391	0.9863	10.0874	58.0356	16.5102	1.1156	3.2459
	%XPT	2.4505	97.5547	99.4982	1.3927	79.4987	76.609	99.0137	89.9126	41.9644	83.4898	98.8844	96.7541
	AB	4.22E+01	4.88E+01	**3.00E+01**	5.84E+01	3.00E+01	**3.00E+01**	**3.00E+01**	3.37E+01	3.47E+01	3.06E+01	**3.00E+01**	**3.00E+01**
F12	**MD**	4.20E+01	4.80E+01	**3.00E+01**	5.90E+01	3.00E+01	**3.00E+01**	**3.00E+01**	3.50E+01	3.50E+01	3.00E+01	**3.00E+01**	**3.00E+01**
	SD	1.63E+00	3.54E+00	**0.00E+00**	2.91E+00	0.00E+00	**0.00E+00**	**0.00E+00**	3.70E+00	4.15E+00	7.74E-01	1.83E-01	1.83E-01
	%XPL	56.3797	2.2746	4.6431	98.0768	16.0563	17.8066	3.5525	9.1007	39.3777	27.7303	4.3927	4.7957
	%XPT	43.6203	97.7254	95.3569	1.9232	83.9437	82.1934	96.4475	90.8993	60.6223	72.2697	95.6073	95.2043
	AB	2.75E+03	1.06E-01	8.01E-05	1.93E+03	8.35E+01	1.31E+00	1.46E-06	4.56E+02	1.36E+03	2.96E+00	8.93E-06	**7.44E-07**
F13	**MD**	2.88E+03	1.36E-02	7.46E-05	1.92E+03	8.25E+01	1.03E+00	3.72E-07	1.88E+02	5.62E+02	2.66E+00	1.05E-06	**1.88E-08**
	SD	7.50E+02	3.94E-01	3.69E-05	3.15E+02	2.08E+01	9.05E-01	2.44E-06	6.44E+02	1.48E+03	1.29E+00	1.75E-05	**1.85E-06**
	%XPL	90.316	2.4176	1.0526	98.9323	70.1808	27.7344	1.1843	6.6424	22.9571	31.7263	1.1151	2.8662
	%XPT	9.684	97.5824	98.9474	1.0677	29.8192	72.2656	98.8157	93.3576	77.0429	68.2737	98.8849	97.1338
	AB	4.28E+10	4.38E+09	**2.29E-28**	2.08E+10	6.42E+02	1.34E+04	1.22E+03	2.08E+09	1.16E+02	7.98E+00	1.08E-05	2.24E+02
F14	**MD**	4.40E+10	4.08E+09	**2.14E-28**	2.03E+10	6.66E+02	1.16E+04	1.17E+03	2.90E-01	5.59E+01	7.04E+00	7.41E-07	1.61E+02
	SD	9.36E+09	2.48E+09	**6.23E-29**	4.80E+09	7.47E+01	5.72E+03	5.86E+02	1.14E+10	1.66E+02	2.86E+00	3.56E-05	2.32E+02
	%XPL	93.7338	4.5958	2.0125	98.6879	12.5087	15.4941	0.7743	8.0601	42.6197	9.9536	1.014	2.1986
	%XPT	6.2662	95.4042	97.9875	1.3121	87.4913	84.5059	99.2257	91.9399	57.3803	90.0464	98.986	97.8014
	AB	3.57E+01	1.31E+01	9.61E+00	2.43E+01	1.10E+01	9.92E+00	8.74E+00	1.19E+01	1.49E+01	1.16E+01	**8.78E+00**	9.10E+00
F15	**MD**	4.69E+04	5.34E+01	**2.25E-09**	1.40E+04	1.35E+01	2.16E+01	1.46E+01	7.41E+03	2.23E-01	1.98E+01	1.52E-02	1.16E+01

		ABC	BA	CMAES	CSA	DE	FA	GWO	MFO	PSO	SSO	TLBO	WOA
	SD	4.70E+04	5.32E+01	**6.88E-15**	1.32E+04	1.32E+01	2.16E+01	1.42E+01	4.72E-03	1.57E-01	1.97E+01	3.32E-04	1.14E+01
	%XPL	94.4582	2.9616	11.8761	98.0231	37.4339	28.9769	3.3576	10.4171	49.7007	51.1412	17.15	7.1632
	%XPT	5.5418	97.0384	88.1239	1.9769	62.5661	71.0231	96.6424	89.5829	50.2993	48.8588	82.85	92.8368
	AB	1.24E+04	1.60E+01	**1.23E-08**	3.76E+03	1.28E+00	2.31E+00	4.90E+00	2.26E+04	2.16E-01	2.69E+00	5.52E-02	5.55E+00
F16	**MD**	2.97E+02	6.57E+01	7.02E+01	2.58E+02	8.34E+01	5.90E+01	1.52E-01	1.36E+02	1.38E+02	5.71E+01	1.27E+01	**3.79E-15**
	SD	2.97E+02	6.17E+01	5.27E+01	2.65E+02	8.40E+01	5.45E+01	0.00E+00	1.25E+02	1.26E+02	5.72E+01	1.29E+01	**0.00E+00**
	%XPL	95.3699	4.8089	2.1825	98.2647	18.5119	17.4896	2.4108	5.9925	46.623	19.3011	1.1184	5.3582
	%XPT	4.6301	95.1911	97.8175	1.7353	81.4881	82.5104	97.5892	94.0075	53.377	80.6989	98.8816	94.6418
	AB	1.61E+01	3.21E+01	4.92E+01	1.60E+01	6.98E+00	1.91E+01	8.35E-01	3.52E+01	4.59E+01	1.21E+01	4.78E+00	**1.44E-14**
F17	**MD**	2.38E+05	4.60E+02	1.13E+01	2.41E+05	7.54E+01	8.43E+01	2.70E+01	5.71E+04	8.45E+04	9.00E+01	2.31E+01	**2.64E+01**
	SD	2.46E+05	1.74E+02	1.59E-01	2.49E+05	7.51E+01	8.61E+01	2.71E+01	5.62E+04	5.81E+04	9.59E+01	2.31E+01	**2.63E+01**
	%XPL	89.3263	4.9894	4.2105	98.3033	46.8273	26.7953	1.3108	3.1251	50.282	24.4155	5.9168	2.3435
	%XPT	10.6737	95.0106	95.7895	1.6967	53.1727	73.2047	98.6892	96.8749	49.718	75.5845	94.0832	97.6565
	AB	6.46E+04	7.57E+02	3.19E+01	7.70E+04	2.21E+01	4.21E+01	7.33E-01	4.50E+04	5.50E+04	4.90E+01	7.63E-01	**3.01E-01**
F18	**MD**	6.67E+01	3.29E+01	1.64E-13	4.97E+01	1.45E+01	9.38E+00	3.88E-25	5.29E+01	2.21E+01	1.70E-01	**4.54E-35**	3.35E+01
	SD	6.77E+01	3.31E+01	1.38E-13	5.08E+01	1.48E+01	9.60E+00	1.85E-25	5.07E+01	2.12E+01	1.65E-01	**4.03E-35**	2.80E+01
	%XPL	95.3792	3.5059	2.4252	98.6945	20.3476	18.411	0.7932	6.3083	39.2793	23.6444	1.0447	1.8732
	%XPT	4.6208	96.4941	97.5748	1.3055	79.6524	81.589	99.2068	93.6917	60.7207	76.3556	98.9553	98.1268
	AB	4.91E+00	4.22E+00	1.09E-13	3.43E+00	1.49E+00	2.33E+00	4.73E-25	1.14E+01	5.09E+00	2.72E-02	**2.75E-35**	3.05E+01
F19	**MD**	2.59E+32	7.27E+27	1.03E+01	7.60E+29	3.11E-02	1.81E+01	2.41E-57	4.73E+02	7.42E+02	7.35E+04	4.36E-42	**2.80E-168**
	SD	2.49E+31	3.39E+22	1.23E-14	1.30E+29	3.02E-02	1.72E+01	5.61E-58	5.00E+02	8.00E+02	2.57E+02	3.38E-42	**5.92E-178**
	%XPL	96.9787	1.7034	2.387	98.617	48.5372	22.2809	0.7828	13.8845	64.5812	5.2652	1.1583	17.2994
	%XPT	3.0213	98.2966	97.613	1.383	51.4628	77.7191	99.2172	86.1155	35.4188	94.7348	98.8417	82.7006
	AB	6.87E+32	3.95E+28	5.66E+01	1.61E+30	6.65E-03	3.87E+00	4.40E-57	1.66E+02	2.06E+02	3.44E+05	2.93E-42	**0.00E+00**
F20	**MD**	-4.98E+49	-1.43E+268	-5.82E+27	-8.05E+03	-1.21E+04	-9.12E+03	**-5.91E+03**	-8.56E+03	-8.21E+03	-7.28E+03	-7.40E+03	-1.18E+04
	SD	-1.02E+48	-3.81E+260	-3.35E+23	-7.81E+03	-1.22E+04	-8.99E+03	**-5.96E+03**	-8.35E+03	-8.04E+03	-7.21E+03	-7.76E+03	-1.23E+04
	%XPL	89.1435	1.5237	3.685	98.1749	12.5994	20.5564	0.5686	5.5311	38.7221	19.2984	1.386	2.6828
	%XPT	10.8565	98.4763	96.315	1.8251	87.4006	79.4436	99.4314	94.4689	61.2779	80.7016	98.614	97.3172
	AB	1.94E+50	6.55E+04	2.90E+28	8.68E+02	**4.21E+02**	6.92E+02	7.23E+02	9.93E+02	1.01E+03	5.62E+02	1.13E+03	1.18E+03
F21	**MD**	1.49E+04	8.74E+03	0.00E+00	1.91E+04	**0.00E+00**	1.39E+01	0.00E+00	1.67E+03	1.33E+03	2.00E-01	0.00E+00	1.00E-01
	SD	1.47E+04	8.37E+03	0.00E+00	1.90E+04	**0.00E+00**	1.40E+01	0.00E+00	0.00E+00	0.00E+00	0.00E+00	0.00E+00	0.00E+00
	%XPL	1.12	0.1531	0.8569	99.1186	83.7028	25.9312	19.3947	4.2543	8.3847	9.2638	43.7278	12.846
	%XPT	98.88	99.8469	99.1431	0.8814	16.2972	74.0688	80.6053	95.7457	91.6153	90.7362	56.2722	87.154
	AB	2.27E+03	2.54E+03	0.00E+00	3.32E+03	0.00E+00	2.35E+00	0.00E+00	4.61E+03	3.46E+03	4.07E-01	**0.00E+00**	3.05E-01
F22	**MD**	-6.58E+02	-9.93E+02	-1.04E+03	-7.24E+02	-1.17E+03	-1.07E+03	-9.31E+02	-1.02E+03	-1.08E+03	-1.07E+03	**-1.03E+03**	-1.17E+03
	SD	-6.59E+02	-9.91E+02	-1.04E+03	-7.18E+02	-1.17E+03	-1.08E+03	-9.47E+02	-1.02E+03	-1.08E+03	-1.07E+03	**-1.02E+03**	-1.17E+03
	%XPL	77.4735	1.9342	2.1311	98.3742	9.9359	15.8333	0.6335	6.08	42.6055	6.2696	1.2032	2.3062
	%XPT	22.5265	98.0658	97.8689	1.6258	90.0641	84.1667	99.3665	93.92	57.3945	93.7304	98.7968	97.6938
	AB	3.50E+01	3.56E+01	3.29E+01	2.40E+01	2.56E-03	4.08E+01	6.07E+01	4.52E+01	2.81E+01	2.81E+01	2.59E+01	**2.03E+01**
F23	**MD**	1.36E+06	5.66E+05	-4.93E+03	1.19E+06	4.34E+05	9.24E+03	**-2.50E+02**	2.60E+05	7.59E+05	-3.10E+03	-1.02E+03	-3.77E+03
	SD	1.41E+06	5.20E+05	-4.93E+03	1.17E+06	4.22E+05	7.22E+03	**-2.10E+02**	1.22E+05	6.82E+05	-3.21E+03	-1.02E+03	-3.79E+03
	%XPL	96.0793	7.0659	2.1461	99.2405	30.4547	20.9524	20.571	4.8452	42.127	26.2275	4.9875	16.4037
	%XPT	3.9207	92.9341	97.8539	0.7595	69.5453	79.0476	79.429	95.1548	57.873	73.7725	95.0125	83.5963
	AB	1.89E+05	2.13E+05	**5.24E-11**	1.72E+05	6.51E+04	7.72E+03	1.28E+02	2.79E+05	3.28E+05	3.52E+02	1.46E+02	2.91E+02
F24	**MD**	4.69E+04	5.34E+01	**2.25E-09**	1.40E+04	1.35E+01	2.16E+01	1.46E+01	7.41E+03	2.23E-01	1.98E+01	1.52E-02	1.16E+01
	SD	4.70E+04	5.32E+01	**6.88E-15**	1.32E+04	1.32E+01	2.16E+01	1.42E+01	4.72E-03	1.57E-01	1.97E+01	3.32E-04	1.14E+01
	%XPL	88.4427	5.6739	2.4577	97.6455	73.6445	17.4556	0.739	4.6445	33.1138	4.527	1.0758	5.6605
	%XPT	11.5573	94.3261	97.5423	2.3545	26.3555	82.5444	99.261	95.3555	66.8862	95.473	98.9242	94.3395

Table 6.4: Statistical analysis results for multimodal functions using Friedman test.

Algorithm	Rank
ABC	11.33
BA	8.92
CMAES	**3.58**
CSA	10.96
DE	4.69
FA	6.14
GWO	4.46
MFO	7.46
PSO	7.31
SSO	6.04
TLBO	**3.48**
WOA	**3.62**
P-value	**6.94E-28**

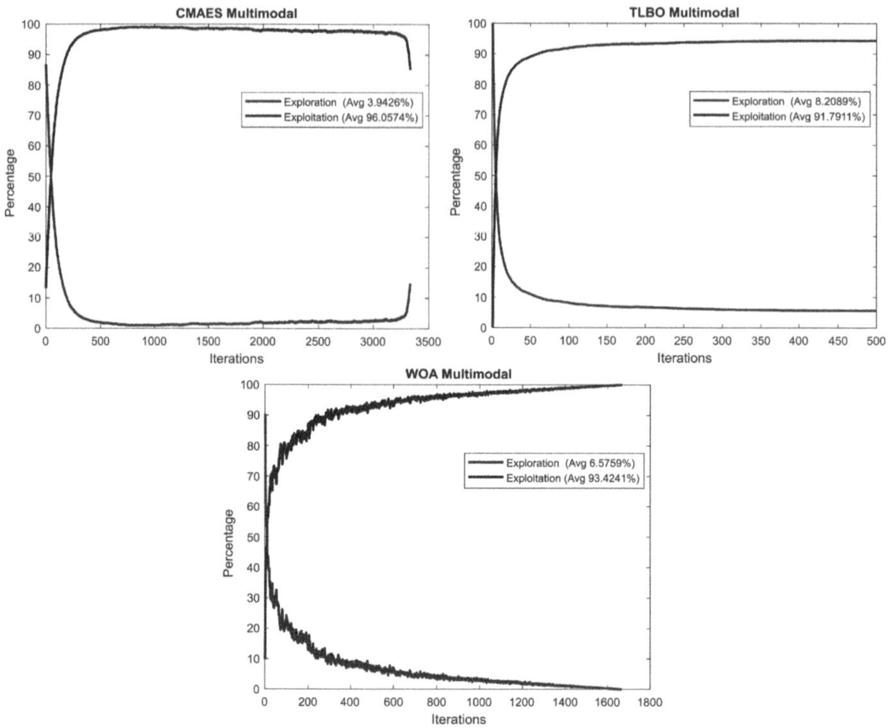

Figure 6.2: Average balance employed by the top three algorithms in multimodal functions.

Tables 6.3 and 6.4, in conjunction with Fig. 6.2, illustrate that CMA-ES outperforms others in four functions with 96.0574% exploitation and 3.9426% exploration. On the other hand, TLBO leads in five functions and ties in three, averaging 91.7911% exploitation and 8.2089% exploration. Lastly, WOA excels in seven functions and ties in four, balancing 93.4241% exploitation and 6.5759% exploration. Based on these results, which indicate a brief exploration and prolonged exploitation, optimal performance for multimodal functions suggests a balance near 90% exploitation and 10% exploration. CMA-ES and WOA's non-greedy selection mechanisms and TLBO's greedy selection influence these rates, with attraction operators enhancing exploitation.

The Friedman test, applied to evaluate the overall performance across multimodal functions, indicates a statistically significant difference among the algorithms ($p<0.001$). The ranking derived from the Friedman test places CMAES at the top, followed closely by TLBO and WOA.

Thanks to the dimension-wise diversity metric, the tables and figures in this section provide a comprehensive analysis of the performance of various algorithms on multimodal functions. The results highlight the strengths and weaknesses of each algorithm, offering valuable insights for researchers and practitioners in the field of optimization.

6.3.1.3 Unimodal Functions

This subsection focuses on the results for unimodal functions, characterized by a single global optimum. Table 6.5 compares the performance of the selected algorithms on these functions, while Table 6.6 presents the statistical analysis results using the Friedman test. The best results are highlighted in bold. The analysis suggests that the best performing algorithms are GWO, TLBO, and WOA.

Table 6.5: Best solutions found and balance employed in each unimodal function.

Function		ABC	BA	CMAES	CSA	DE	FA	GWO	MFO	PSO	SSO	TLBO	WOA
	AB	7.84E+04	3.67E+04	9.47E-37	1.18E+05	1.21E-03	7.10E+01	4.02E-100	1.23E+04	3.38E+04	1.88E+00	7.60E-86	**2.36E-250**
F25	**MD**	7.85E+04	3.33E+04	5.29E-37	1.16E+05	1.20E-03	6.88E+01	4.98E-101	4.29E+03	3.01E+04	1.70E+00	4.10E-86	**7.96E-265**
	SD	9.04E+03	1.42E+04	9.32E-37	1.46E+04	2.75E-04	1.17E+01	1.54E-99	1.87E+04	2.39E+04	7.23E-01	9.83E-86	**0.00E+00**
	%XPL	79.9258	2.0399	2.2269	98.8173	10.1382	16.6425	0.4699	6.796	39.3438	7.185	1.0403	1.7461
	%XPT	20.0742	97.9601	97.7731	1.1827	89.8618	83.3575	99.5301	93.204	60.6562	92.815	98.9597	98.2539
	AB	2.85E+06	1.59E+06	5.43E-34	3.99E+06	2.75E-02	2.96E+03	4.42E-98	1.09E+06	1.69E+06	2.21E+02	4.32E-84	**7.05E-251**
F26	**MD**	2.85E+06	1.36E+06	2.32E-34	4.00E+06	2.77E-02	2.97E+03	1.24E-99	1.09E+06	1.57E+06	1.58E+02	2.15E-84	**1.51E-264**
	SD	4.42E+05	7.66E+05	8.08E-34	5.07E+05	7.20E-03	4.48E+02	1.00E-97	9.16E+05	8.83E+05	1.93E+02	9.66E-84	**0.00E+00**
	%XPL	89.6177	3.6873	2.3025	98.3686	11.9995	19.2656	0.5263	5.9581	37.2827	6.0095	1.0183	1.6362
	%XPT	10.3823	96.3127	97.6975	1.6314	88.0005	80.7344	99.4737	94.0419	62.7173	93.9905	98.9817	98.3638
	AB	3.86E+01	1.54E-05	2.45E-43	5.12E+01	6.33E-07	3.14E-02	2.25E-103	6.99E+00	1.75E+00	9.50E-02	5.25E-89	**4.47E-250**
F27	**MD**	4.03E+01	1.54E-05	1.96E-43	5.13E+01	5.77E-07	3.14E-02	1.74E-104	1.52E-07	2.13E-06	9.43E-02	2.49E-89	**1.08E-267**
	SD	6.69E+00	2.27E-06	1.81E-43	7.05E+00	2.10E-07	3.60E-03	4.39E-103	1.53E+01	6.65E+00	1.43E-02	7.54E-89	**0.00E+00**
	%XPL	77.6193	1.8221	1.9656	98.5818	9.6177	14.848	0.4756	5.5791	43.1609	26.0592	0.9471	1.7297
	%XPT	22.3807	98.1779	98.0344	1.4182	90.3823	85.152	99.5244	94.4209	56.8391	73.9408	99.0529	98.2703
	AB	1.85E+03	1.16E+01	9.94E-41	2.47E+03	2.85E-05	1.58E+00	9.20E-102	4.43E+02	8.23E+02	1.94E+00	3.41E-87	**1.25E-253**
F28	**MD**	1.80E+03	4.09E+00	6.99E-41	2.54E+03	2.73E-05	1.59E+00	8.48E-103	2.50E+02	8.00E+02	1.97E+00	1.40E-87	**2.19E-267**
	SD	2.54E+02	1.70E+01	9.10E-41	3.71E+02	1.06E-05	1.95E-01	3.68E-101	5.50E+02	5.24E+02	4.85E-01	4.08E-87	**0.00E+00**
	%XPL	81.7367	3.4067	2.1688	98.4859	10.7318	16.4721	0.4746	5.8278	40.2838	16.9878	0.9162	1.9663
	%XPT	18.2633	96.5933	97.8312	1.5141	89.2682	83.5279	99.5254	94.1722	59.7162	83.0122	99.0838	98.0337
	AB	7.55E-02	1.30E-08	5.08E-09	7.70E-03	1.31E-22	6.31E-08	**0.00E+00**	8.48E-19	3.51E-13	3.26E-05	2.44E-197	**0.00E+00**
F29	**MD**	7.33E-02	1.12E-08	4.28E-09	7.86E-03	3.35E-23	5.47E-08	**0.00E+00**	2.90E-23	6.81E-16	3.22E-05	5.98E-199	**0.00E+00**
	SD	3.51E-02	8.50E-09	2.89E-09	4.56E-03	3.62E-22	4.56E-08	**0.00E+00**	4.19E-18	1.88E-12	1.79E-05	0.00E+00	**0.00E+00**
	%XPL	95.81	5.5193	0.2584	98.8009	27.213	27.8819	0.9651	11.4744	52.5785	54.9875	1.5114	2.2625
	%XPT	4.19	94.4807	99.7416	1.1991	72.787	72.1181	99.0349	88.5256	47.4215	45.0125	98.4886	97.7375

WOA emerges as the most effective algorithm for unimodal functions, with an average rank of 1. Figure 6.3 compares the performance of the top three algorithms (WOA, TLBO, and GWO). All three algorithms predominantly exploited the search space, with WOA exploiting 98.1318% of the time, TLBO 98.9133%, and GWO 99.4177%.

WOA and GWO both utilize non-greedy selection mechanisms, and even though they differ in the number of attraction operators, they operate in a similar manner. WOA has a single attraction operator driven by the best solution, while GWO uses an average of the top three best solutions. Ultimately, both direct the search towards the area with the best-found solution. On the other hand, TLBO has two distinct phases: one where every search agent moves towards a "Teacher," representing the best-found solution, and a "Learning" phase, where the movements are random. At first glance, this seems dissimilar to the behavior of the WOA and GWO algorithms, but once the greedy selection mechanism takes effect, the movements towards the Teacher (best-found solution) are the most likely to be accepted. This similarity in their mechanisms leads to comparable performance in terms of solution quality and exploration-exploitation balance, as illustrated in Fig. 6.3.

Table 6.6: Statistical analysis results for unimodal functions using Friedman test.

Algorithm	Rank
ABC	11.20
BA	8.20
CMAES	4.60
CSA	11.80
DE	4.80
FA	7.20
GWO	**2**
MFO	8
PSO	8.80
SSO	7.40
TLBO	**3**
WOA	**1**
P-value	**5.20E-7**

6.3.1.4 Hybrid Functions

Hybrid functions combine features from both unimodal and multimodal functions, creating complex landscapes that challenge the adaptability and robustness of optimization algorithms. Table 6.7 compares the performance of selected algorithms on these hybrid functions, while Table 6.8 presents the statistical analysis outcomes using the Friedman test. The best results are highlighted in boldface. Analyzing these tables reveals that WOA and TLBO consistently performed the best, with CMA-ES following at a distant third.

Table 6.7: Best solutions found and balance employed in each hybrid function.

Function		ABC	BA	CMAES	CSA	DE	FA	GWO	MFO	PSO	SSO	TLBO	WOA
	AB	8.57E+04	1.12E+04	3.68E-19	1.99E+04	4.42E-03	1.95E+01	2.39E-58	2.01E+04	2.48E+04	2.92E-01	1.35E-43	**2.33E-175**
F30	**MD**	7.82E+04	8.84E+03	2.45E-19	2.05E+04	4.38E-03	1.98E+01	1.85E-58	2.01E+04	2.01E+04	2.88E-01	1.16E-43	**7.02E-181**
	SD	2.77E+04	7.55E+03	4.81E-19	2.33E+03	7.50E-04	1.94E+00	2.24E-58	1.86E+04	1.53E+04	4.99E-02	9.99E-44	**0.00E+00**
	%XPL	84.7161	4.8273	3.4316	98.5073	10.4171	16.2454	0.4767	9.1888	44.0224	8.2539	1.0883	1.8183
	%XPT	15.2839	95.1727	96.5684	1.4927	89.5829	83.7546	99.5233	90.8112	55.9776	91.7461	98.9117	98.1817
	AB	7.98E+02	4.00E+02	8.22E+01	7.21E+02	2.90E+01	7.98E+01	2.90E+01	1.80E+02	1.55E+02	7.31E+01	**2.90E+01**	**2.90E+01**
F31	**MD**	7.98E+02	4.03E+02	7.78E+01	7.18E+02	2.90E+01	7.61E+01	2.90E+01	1.18E+02	8.95E+01	7.64E+01	**2.90E+01**	**2.90E+01**
	SD	8.92E+01	6.50E+01	1.84E+01	6.35E+01	2.28E-02	2.05E+01	5.66E-05	1.27E+02	1.26E+02	1.63E+01	**3.30E-12**	7.57E-05
	%XPL	88.0123	2.9059	2.2306	98.6365	15.2593	18.938	0.5138	5.8204	42.3145	7.3259	1.1961	2.5734
	%XPT	11.9877	97.0941	97.7694	1.3635	84.7407	81.062	99.4862	94.1796	57.6855	92.6741	98.8039	97.4266
	AB	2.23E+08	2.62E+06	**3.20E+01**	7.13E+07	4.76E+02	2.45E+02	3.20E+01	7.57E+01	2.40E+02	3.46E+01	**3.20E+01**	7.18E+01
F32	**MD**	2.19E+08	1.50E+06	**3.20E+01**	7.08E+07	4.50E+02	2.45E+02	3.20E+01	7.18E+01	2.31E+02	3.45E+01	**3.20E+01**	6.77E+01
	SD	5.51E+07	2.66E+06	**1.73E-14**	1.92E+07	7.16E+01	4.54E+01	7.26E-06	1.59E+01	7.26E+01	8.49E-01	2.09E-10	3.31E+01
	%XPL	94.507	3.4629	2.4833	98.4313	20.8563	18.7752	0.5362	7.1578	46.595	11.0359	1.2826	4.9243
	%XPT	5.493	96.5371	97.5167	1.5687	79.1437	81.2248	99.4638	92.8422	53.405	88.9641	98.7174	95.0757
	AB	5.54E+03	5.84E+02	7.82E+01	9.21E+02	2.90E+01	5.95E+01	**2.90E+01**	8.95E+02	9.92E+02	1.63E+02	**2.90E+01**	**2.90E+01**
F33	**MD**	5.51E+03	4.89E+02	7.73E+01	9.25E+02	2.90E+01	5.16E+01	**2.90E+01**	8.96E+02	9.24E+02	1.58E+02	**2.90E+01**	**2.90E+01**
	SD	3.22E+03	3.15E+02	2.35E+01	1.13E+02	9.70E-03	2.50E+01	7.14E-15	5.67E+02	5.46E+02	4.43E+01	4.12E-15	**2.64E-15**
	%XPL	90.3856	5.077	3.6581	98.5816	12.9629	17.6859	0.5115	5.5055	46.0492	10.8563	1.2409	1.5329
	%XPT	9.6144	94.923	96.3419	1.4184	87.0371	82.3141	99.4885	94.4945	53.9508	89.1437	98.7591	98.4671

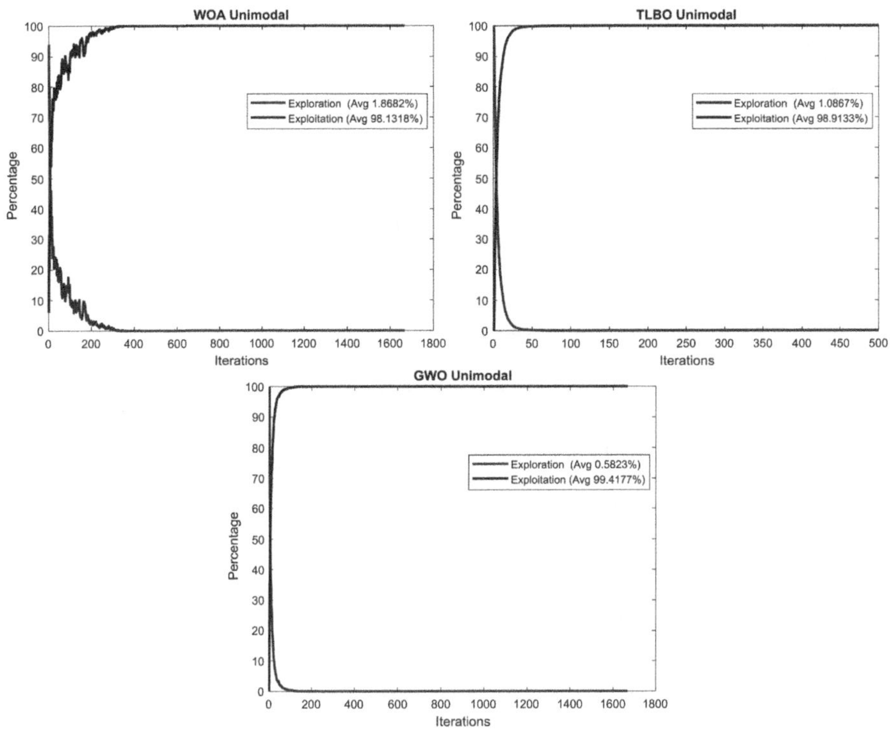

Figure 6.3: Average balance employed by the top three algorithms in unimodal functions.

Table 6.8: Statistical analysis results for hybrid functions using Friedman test.

Algorithm	Rank
ABC	12
BA	9
CMAES	4.50
CSA	10.25
DE	5.50
FA	6.50
GWO	**2.62**
MFO	8.5
PSO	9.25
SSO	5.50
TLBO	**2.12**
WOA	**2.25**
P-value	**1.12E-4**

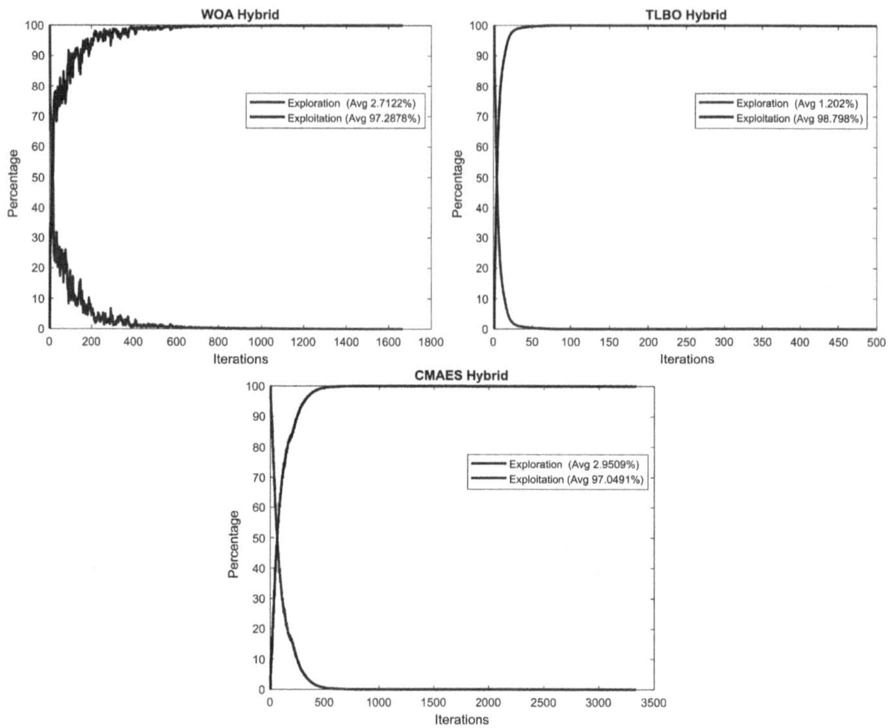

Figure 6.4: Average balance employed by the top three algorithms in hybrid functions.

The results from the aforementioned trio of algorithms are averaged and displayed in Fig. 6.4. The algorithms show consistency in balance employed when compared with previous tests. All three schemes spent above 90% of the runtime exploiting the search space. In almost all hybrid functions the WOA and TLBO algorithms managed to find the global optimum. The only significant difference can be seen in function $_{3}0$ where WOA surpass the other algorithms by a large margin, and $F_{3}2$ where CMA-ES found the best solutions with the least standard deviation. As previously mentioned, the reason for this difference is the specific search mechanisms used for exploration and exploitation. It is important to point out that the WOA algorithm once again produces a rough balance response.

6.3.1.5 Shifted Functions

To explore how algorithms perform when global optima are not near the origin (0,0), shifted functions are employed. These benchmark functions challenge algorithms to adapt their search strategies effectively.

Table 6.9: Best solutions found and balance employed in each shifted function.

Function		ABC	BA	CMAES	CSA	DE	FA	GWO	MFO	PSO	SSO	TLBO	WOA
	AB	1.73E+01	1.31E+01	2.10E+00	1.73E+01	**5.06E-03**	1.82E+00	7.18E+00	7.00E+00	7.51E+00	4.06E-01	4.74E+00	5.43E+00
F34	**MD**	1.74E+01	1.31E+01	**7.11E-15**	1.74E+01	4.65E-03	1.83E+00	8.29E+00	8.29E-01	7.25E-01	3.60E-01	4.55E+00	5.10E+00
	SD	4.73E-01	8.91E-01	6.40E+00	5.12E-01	**1.64E-03**	1.24E-01	3.64E+00	8.42E+00	8.29E+00	1.87E-01	9.31E-01	2.07E+00
	%XPL	87.169	2.6385	2.0468	98.0358	12.9051	15.4772	13.5003	5.4988	43.1621	8.1621	1.2565	16.0797
	%XPT	12.831	97.3615	97.9532	1.9642	87.0949	84.5228	86.4997	94.5012	56.8379	91.8379	98.7435	83.9203
	AB	2.87E+03	1.58E-02	**7.69E-05**	1.86E+03	9.25E+01	1.21E+00	2.20E+01	6.80E+02	1.52E+03	3.05E+00	1.09E-02	1.64E+00
F35	**MD**	2.81E+03	1.11E-02	**7.60E-05**	1.81E+03	9.12E+01	1.05E+00	1.72E+01	2.22E+02	4.84E+02	2.75E+00	8.48E-03	1.24E+00
	SD	6.08E+02	1.64E-02	**2.92E-05**	4.86E+02	2.55E+01	7.00E-01	1.43E+01	1.03E+03	1.60E+03	1.40E+00	7.73E-03	1.23E+00
	%XPL	92.0951	6.081	0.7667	98.9009	64.1413	23.188	29.9811	5.013	26.4855	30.5526	1.6696	26.066
	%XPT	7.9049	93.919	99.2333	1.0991	35.8587	76.812	70.0189	94.987	73.5145	69.4474	98.3304	73.934
	AB	3.01E+02	6.01E+01	7.19E+01	2.61E+02	8.34E+01	5.82E+01	1.30E+02	1.40E+02	1.35E+02	5.89E+01	**1.76E+01**	4.10E+01
F36	**MD**	3.01E+02	5.47E+01	5.22E+01	2.62E+02	8.24E+01	5.21E+01	8.18E+01	1.35E+02	1.38E+02	5.87E+01	**1.71E+01**	1.75E+01
	SD	1.26E+01	2.63E+01	5.23E+01	1.41E+01	8.56E+00	1.93E+01	8.68E+01	2.96E+01	3.58E+01	1.24E+01	**4.51E+00**	5.01E+01
	%XPL	89.7716	1.3032	3.9072	97.923	45.4431	24.725	47.5097	3.2421	40.8761	23.8254	5.7889	27.1777
	%XPT	10.2284	98.6968	96.0928	2.077	54.5569	75.275	52.4903	96.7579	59.1239	76.1746	94.2111	72.8223
	AB	3.32E+05	3.07E+03	**7.77E-01**	5.08E+05	7.05E+01	7.15E+01	1.96E+02	6.00E+04	1.53E+05	1.11E+02	2.80E+01	1.35E+00
F37	**MD**	3.43E+05	1.39E+03	**1.11E-01**	5.24E+05	7.02E+01	5.27E+01	1.48E+02	5.48E+04	1.63E+05	1.09E+02	2.62E+01	7.87E-01
	SD	8.43E+04	3.75E+03	**2.85E+00**	1.13E+05	1.82E+01	4.03E+01	1.65E+02	4.06E+04	6.35E+04	6.34E+01	8.04E+00	1.59E+00
	%XPL	93.895	7.8154	2.7508	98.7681	19.1539	19.5105	48.0801	6.0512	19.8387	24.8886	1.6766	24.1191
	%XPT	6.105	92.1846	97.2492	1.2319	80.8461	80.4895	51.9199	93.9488	80.1613	75.1114	98.3234	75.8809
	AB	7.82E+04	3.48E+04	**1.13E-27**	1.14E+05	1.12E-03	7.44E+01	7.44E+03	1.85E+04	3.41E+04	1.90E+00	6.36E-06	1.10E+02
F38	**MD**	7.78E+04	3.24E+04	**1.21E-27**	1.15E+05	1.02E-03	7.59E+01	5.41E+03	8.59E+03	3.01E+04	1.94E+00	6.51E-07	9.19E+01
	SD	1.20E+04	1.14E+04	**2.77E-28**	1.79E+04	4.19E-04	1.11E+01	5.54E+03	2.34E+04	2.24E+04	6.15E-01	2.73E-05	6.93E+01
	%XPL	80.7225	5.2696	2.3067	98.7206	10.4484	16.7563	11.9413	6.4235	39.5787	7.342	1.3278	15.7178
	%XPT	19.2775	94.7304	97.6933	1.2794	89.5516	83.2437	88.0587	93.5765	60.4213	92.658	98.6722	84.2822
	AB	2.73E+06	1.59E+06	**4.33E-26**	4.18E+06	3.15E-02	2.90E+03	7.74E+05	1.60E+06	1.41E+06	2.60E+02	1.47E-04	2.76E+05
F39	**MD**	2.72E+06	1.47E+06	**1.01E-26**	4.21E+06	3.16E-02	3.00E+03	4.45E+05	1.26E+06	1.25E+06	1.50E+02	5.01E-05	1.85E+04
	SD	3.81E+05	5.22E+05	**7.64E-26**	5.55E+05	1.04E-02	4.50E+02	8.60E+05	1.58E+06	1.11E+06	2.78E+02	2.73E-04	1.09E+06
	%XPL	89.2405	2.9254	2.6224	98.435	12.6352	17.9422	31.7209	7.0813	33.1841	7.7982	1.2846	26.3112
	%XPT	10.7595	97.0746	97.3776	1.565	87.3648	82.0578	68.2791	92.9187	66.8159	92.2018	98.7154	73.6888
	AB	2.22E+32	5.11E+29	**7.24E-03**	5.98E+29	3.16E-02	1.84E+01	2.89E+02	4.90E+02	7.83E+02	1.08E+04	5.15E-01	3.45E+19
F40	**MD**	1.72E+31	8.92E+22	**4.62E-14**	1.76E+29	3.02E-02	1.75E+01	2.81E+02	5.00E+02	8.00E+02	3.31E+02	3.88E-03	7.32E+12
	SD	5.78E+32	2.80E+30	**3.97E-02**	8.94E+29	8.41E-03	3.31E+00	6.80E+01	2.07E+02	2.25E+02	4.03E+04	2.69E+00	1.61E+20
	%XPL	87.7802	1.95	3.5449	98.2999	14.0029	23.921	10.3306	6.6679	38.9743	15.4068	1.5227	12.8716
	%XPT	12.2198	98.05	96.4551	1.7001	85.9971	76.079	89.6694	93.3321	61.0257	84.5932	98.4773	87.1284
	AB	3.89E+01	1.52E-05	**7.87E-29**	4.89E+01	5.73E-07	3.17E-02	1.16E-02	2.62E+00	2.62E+00	1.09E-01	2.10E-09	1.07E-02
F41	**MD**	3.84E+01	1.55E-05	**7.57E-29**	4.94E+01	5.70E-07	3.31E-02	1.11E-02	7.83E-08	1.97E-06	1.05E-01	2.22E-10	8.50E-03
	SD	6.80E+00	1.68E-06	**1.47E-29**	5.76E+00	1.44E-07	3.41E-03	3.88E-03	8.00E+00	8.00E+00	2.57E-02	6.54E-09	8.84E-03
	%XPL	79.9896	4.5125	2.1102	98.152	9.6874	16.0075	47.753	6.3734	39.701	28.9529	1.1754	26.0626
	%XPT	20.0104	95.4875	97.8898	1.848	90.3126	83.9925	52.247	93.6266	60.299	71.0471	98.8246	73.9374
	AB	1.74E+03	1.96E+01	**7.64E-28**	2.57E+03	2.79E-05	1.67E+00	1.70E+00	4.13E+02	5.90E+02	1.75E+00	6.04E-07	8.74E-01
F42	**MD**	1.76E+03	9.75E+00	**7.45E-28**	2.54E+03	2.56E-05	1.67E+00	1.36E+00	2.50E+02	5.50E+02	1.80E+00	1.58E-08	6.74E-01
	SD	2.02E+02	2.59E+01	**2.89E-28**	3.58E+02	9.89E-06	2.47E-01	1.16E+00	4.74E+02	3.86E+02	3.80E-01	2.65E-06	8.38E-01
	%XPL	78.9622	2.7921	2.2211	97.9084	10.5863	17.3523	47.1852	5.945	37.1032	18.8607	1.2538	26.1591
	%XPT	21.0378	97.2079	97.7789	2.0916	89.4137	82.6477	52.8148	94.055	62.8968	81.1393	98.7462	73.8409

Table 6.10: Statistical analysis results for shifted functions using Friedman test.

Algorithm	**Rank**
ABC	11.44
BA	7.44
CMAES	**1.89**
CSA	11.55
DE	3.78
FA	4.56
GWO	6.78
MFO	8.67
PSO	9
SSO	5.44
TLBO	**2.44**
WOA	5
P-value	**1.12E-4**

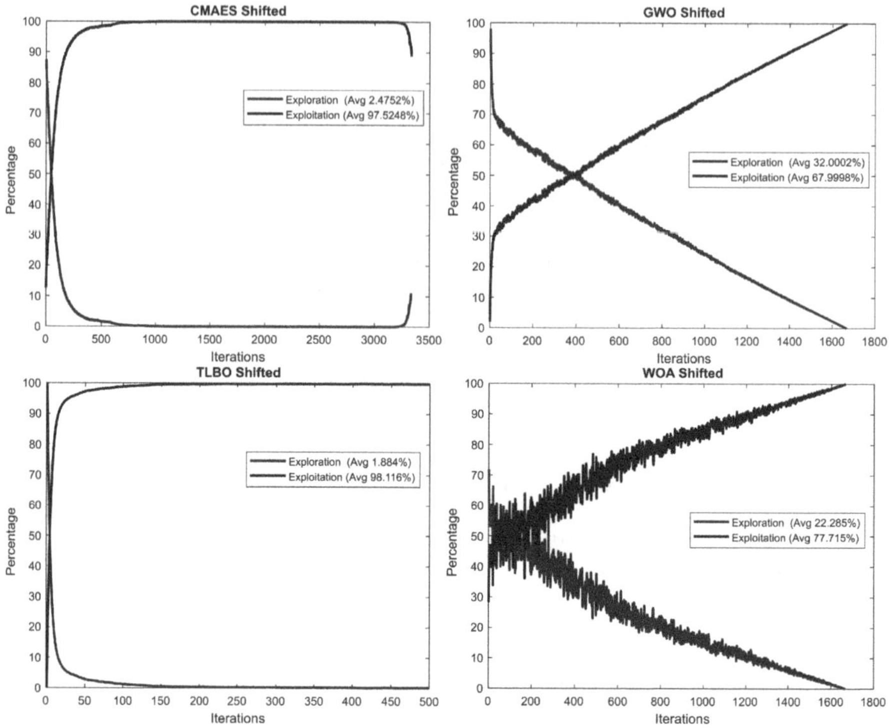

Figure 6.5: Average balance employed by CMA-ES, GWO, TLBO and WOA algorithms in shifted functions.

The results in shifted functions reveal that CMA-ES maintains robust performance across different shifted positions, winning in 7 out of 9 functions with minimal variability. Conversely, algorithms like GWO and WOA struggle significantly in these scenarios, showing prolonged exploration phases that hinder their performance compared to static functions. TLBO maintains a consistent balance between exploration and exploitation, though with slightly reduced solution quality compared to non-shifted cases.

These analyses underscore the critical influence of function positioning on algorithm performance, highlighting strengths and weaknesses in adapting to shifted global optima.

6.3.1.6 Conclusions

The analysis reveals several noteworthy patterns from the experimental results. Across the study, the metaheuristic algorithms WOA, CMA-ES, GWO, and TLBO consistently achieved an exploitation rate exceeding 90% (XPL%=10, XPT%=90). Notably, WOA consistently demonstrated superior performance indices among all evaluated schemes.

The experiments underscored the significance of maintaining a specific balance (XPL%=10, XPT%=90) for optimal performance. This optimal behavior entails an initial phase of exploration primarily within the first 100 to 300 iterations, followed by a predominant exploitation strategy for the remainder of the search process.

An intriguing characteristic observed in the balance graphs is their roughness, indicating erratic fluctuations in balance during evolution. These fluctuations, induced by subtle and abrupt changes in diversity facilitated by the search mechanisms, tend to slightly increase the exploration rate. This phenomenon allows algorithms to escape local minima even during the exploitation phase, contributing to improved overall performance. WOA consistently exhibits a rough balance response (Figs. 6.2, 6.3, 6.4), 6.5) which appears to be a key factor contributing to its robust performance.

Furthermore, the study highlights the crucial role of combining competitive search mechanisms with an appropriate balance response for achieving superior performance. Interestingly, even when two metaheuristic algorithms exhibited similar balance responses, their performance in terms of solution quality varied significantly in some instances. This disparity underscores the importance of maintaining an optimal balance (XPL%=10, XPT%=90) alongwith the ability to generate promising solutions through effective operators, essential for leveraging the observed diversity in the balance response.

In the context of shifted functions, the results present a critical case for analysis, as many algorithms failed to maintain the high performance observed in non-shifted functions. Among the top performers, only CMA-ES managed to maintain competitive performance across both shifted and non-shifted versions of the functions. In contrast, GWO and WOA not only exhibited poor solution quality but also displayed inappropriate balance responses under shifted function conditions. Known for their effectiveness when the optimal point is centered at the origin, these algorithms struggled significantly when faced with shifted optimal points. This challenge led to prolonged exploration phases, disrupting the exploration-exploitation balance and consequently compromising solution quality and precision.

6.3.2 *Evaluating Population Health*

The optimization of metaheuristic algorithms, particularly Particle Swarm Optimization (PSO), has drawn significant attention from researchers aiming to enhance performance through parameter tuning. Traditional approaches often rely on fixed parameter settings, but advancements in soft computing, such as Fuzzy Logic Systems (FLS), have revolutionized how these algorithms can dynamically adapt to complex optimization landscapes.

In seminal works, Shi and Eberhart introduced Fuzzy Adaptive PSO in 2001, pioneering the use of FLS to adjust the inertia weight based on the current best performance and inertia weight value [52]. This early adaptation marked a shift towards more adaptive and responsive PSO variants. Building upon this, dynamic PSO, as proposed by Melin et al., further refined parameter adaptation by incorporating population diversity, iteration count, and swarm-best particle error into FLS controllers for cognitive and social parameters [53]. Similarly, Tian and Yang extended PSO with a learning coefficient controlled by FLS, responding to changes in global optimum increments and particle fitness deviations [54]. These adaptations underscore the utility of FLS in enhancing PSO's robustness across diverse optimization scenarios.

This section explores an advanced iteration of PSO leveraging FLS to govern the inertia weight dynamically. Central to this innovation is the integration of a stagnation metric, alongside dimension-wise diversity metrics [55, 56] to determine population health. By utilizing these inputs, the FLS controller effectively computes an optimal inertia weight. This dynamic approach not only accelerates convergence by mitigating suboptimal trapping but also tailors parameter values to the intricacies of each optimization problem.

6.3.2.1 Fuzzy Logic System Controller

The adaptive control of metaheuristic algorithm parameters, such as the inertia weight (w) in Particle Swarm Optimization (PSO), plays a crucial role in optimizing convergence behavior [57]. Fuzzy Logic Systems (FLS) have emerged as effective tools for dynamically adjusting these parameters based on real-time feedback from the algorithm's performance [58].

The Fuzzy Logic System controller comprises three primary stages:

Fuzzification: Inputs, such as stagnation (measuring the persistence of unchanged fitness values over iterations) and dimension-wise diversity (reflecting the variability across solution dimensions), are transformed into fuzzy sets suitable for fuzzy inference.

Fuzzy Inference: Using the Mamdani algorithm, the fuzzy controller evaluates predefined rules based on the fuzzy sets generated in the fuzzification stage. These rules dictate how inputs relate to the desired output, w, which controls the inertia weight of PSO.

Defuzzification: After inference, which yields fuzzy output sets, defuzzification aggregates these sets into a crisp value for w. This process typically involves centroid calculation or other aggregation methods to convert fuzzy outputs into a precise numerical value suitable for algorithmic parameter adjustment.

Figure 6.6 illustrates the architecture of the proposed fuzzy controller for adjusting w, integrating inputs related to stagnation (Fig. 6.7(a)) and population diversity (Fig. 6.7(b)), with corresponding membership functions. Table 6.11 outlines the specific rules guiding the fuzzy inference process, delineating how different combinations of stagnation and diversity inputs influence the output inertia weight.

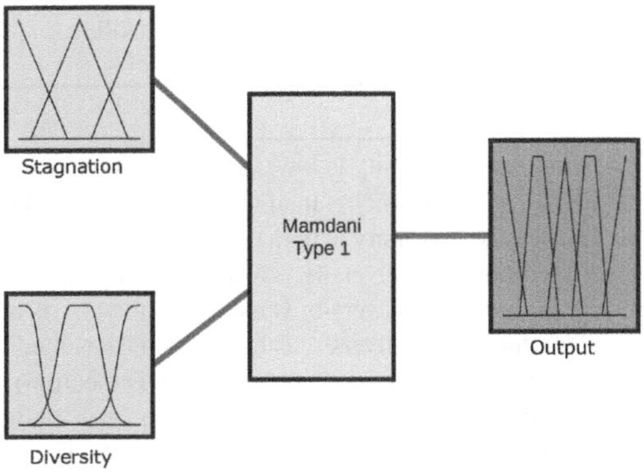

Figure 6.6: Fuzzy logic system controller applied to the PSO algorithm.

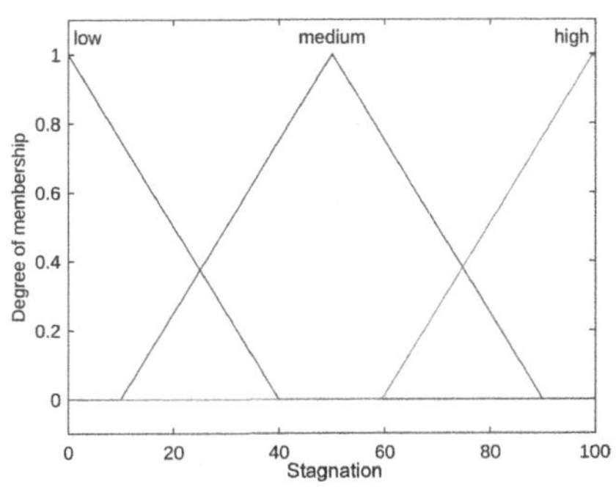

(a) Input related to stagnation of the population.

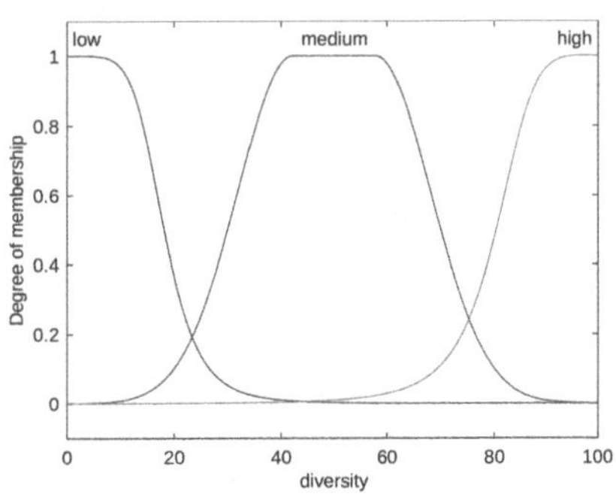

(b) Input related to the diversity of the population.

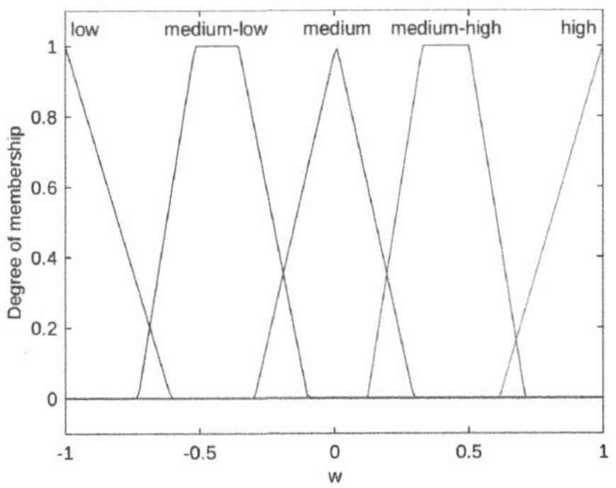

(c) Output related to the Inertia of the algorithm

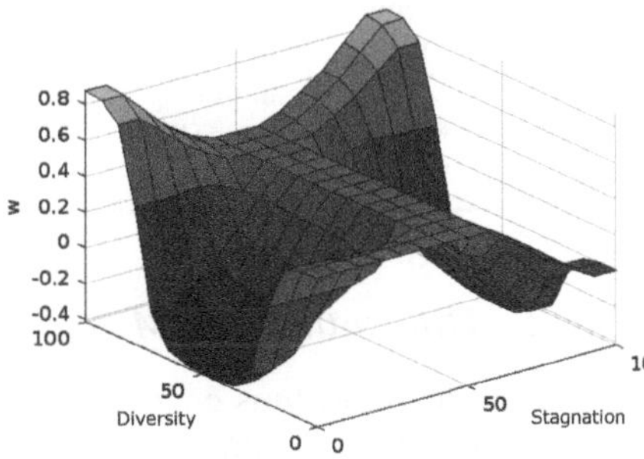

(d) Surface of the fuzzy controller

Figure 6.7: Membership functions and surface of fuzzy controller.

Table 6.11: Rules of the fuzzy controller.

Rules
1. If (stagnation is low) and (diversity is low), then (w is medium-high)
2. If (stagnation is low) and (diversity is medium), then (w is medium-low)
3. If (stagnation is low) and (diversity is high), then (w is high)
4. If (stagnation is medium) and (diversity is low), then (w is medium-high)
5. If (stagnation is medium) and (diversity is medium), then (w is medium-high)
6. If (stagnation is medium) and (diversity is high), then (w is medium-high)
7. If (stagnation is high) and (diversity is low), then (w is medium)
8. If (stagnation is high) and (diversity is medium), then (w is medium-low)
9. If (stagnation is high) and (diversity is high), then (w is high)

Table 6.12: Parameters of the membership function related to the inputs of the fuzzy system.

Input: Stagnation	**Type**	**Parameters**
Function of Membership: Low	Triangular	[-40 0 40]
Function of Membership: Medium	Triangular	[10 50 90]
Function of Membership: High	Triangular	[60 100 140]
Input: Diversity	**Type**	**Parameters**
Function of Membership: Low	Gaussian Bell	[19 3 -1]
Function of Membership: Medium	Gaussian Bell	[10 42 10 57]
Function of Membership: High	Gaussian Bell	[20 2 100]

Table 6.13: Parameters of the membership functions related to the output of the fuzzy system.

Output: W inertia	**Type**	**Parameters**
Fn. Memb.: Low	Triangular	[–1.8 –1 –0.6047]
Fn. Memb.: Medium-Low	Trapezoidal	[–0.728 –0.516 –0.356 –0.0972]
Fn. Memb.: Medium	Triangular	[–0.307 –0.00212 0.285]
Fn. Memb.: Medium-High	Trapezoidal	[0.1226 0.33 0.5 0.71]
Fn. Memb.: High	Triangular	[0.6131 1 1.8]

6.3.2.2 Difficulties with Metaheuristic Algorithms

Metaheuristic algorithms encounter face common issues such as stagnation, local minima entrapment, inefficient function evaluations, and complex parameter tuning. The diverse nature of optimization problems complicates parameter settings—what works well for one problem may hinder performance in another. These challenges are summarized as follows:

- Stagnation and Local Minima Entrapment: A major challenge in metaheuristics is stagnation, where algorithms get stuck in suboptimal solutions. This occurs when premature convergence leads to local minima entrapment, preventing further exploration and obstructing global optimization.

- Ineffective Use of Function Evaluations: Metaheuristic algorithms operate with limited function evaluations, requiring efficient utilization. Inefficiencies arise when excessive exploration iterations fail to yield substantial solution improvements, affecting optimization effectiveness.
- Parameter Tuning Complexity: Optimizing parameters like inertia dampening and velocity clamping demands extensive empirical testing. Achieving the right exploration-exploitation balance is crucial, but without prior knowledge of the problem domain, determining optimal settings remains challenging.

To mitigate these challenges, advanced metrics such as dimension-wise diversity [59], Spearman's rank correlation coefficient [60], and the Ginny index [61] provide valuable insights into the internal dynamics of metaheuristic algorithm populations. These metrics offer objective measures that help algorithm designers create dynamic parameter strategies tailored to each optimization problem's unique requirements.

6.3.2.3 Diversity-based Fuzzy Controlled PSO

In the classic Particle Swarm Optimization (PSO), the inertia damping parameter (w) plays a critical role in balancing exploration and exploitation of the solution space. The choice of w determines whether particles explore widely or exploit local regions for optimal solutions. However, selecting an optimal inertia-damping curve is challenging and problem-dependent. Too rapid a damping can trap the algorithm in local minima, while insufficient damping can lead to erratic movements that hinder convergence.

The proposed PSO variant, referred to as fuzzy-controlled PSO (FCPSO), introduces a dynamic approach to adjust w based on internal feedback gathered during optimization [57]. At each iteration, the algorithm measures the diversity of the population to gauge the spread of search agents across the solution space. This diversity metric informs whether the swarm is in an exploratory phase (agents widely dispersed) or an exploitative phase (agents concentrated around a potential solution).

To detect stagnation, FCPSO compares the current best-found solution with previous iterations. If no improvement is observed over a set number of iterations, a stagnation counter increments, signaling a potential local optimum trap. Together, both metrics allow the proposed algorithm to gauge the health of the current population, or if intervention is required to adjust its parameters.

The fuzzy controller evaluates the current state of the algorithm iteratively. For instance, when diversity indicates high dispersion, implying exploration, the controller adjusts w to maintain a higher acceleration, fostering exploration across the solution space. Conversely, when diversity suggests convergence around a local optimum, indicating exploitation, w is adjusted to lower acceleration to refine the search within the local region.

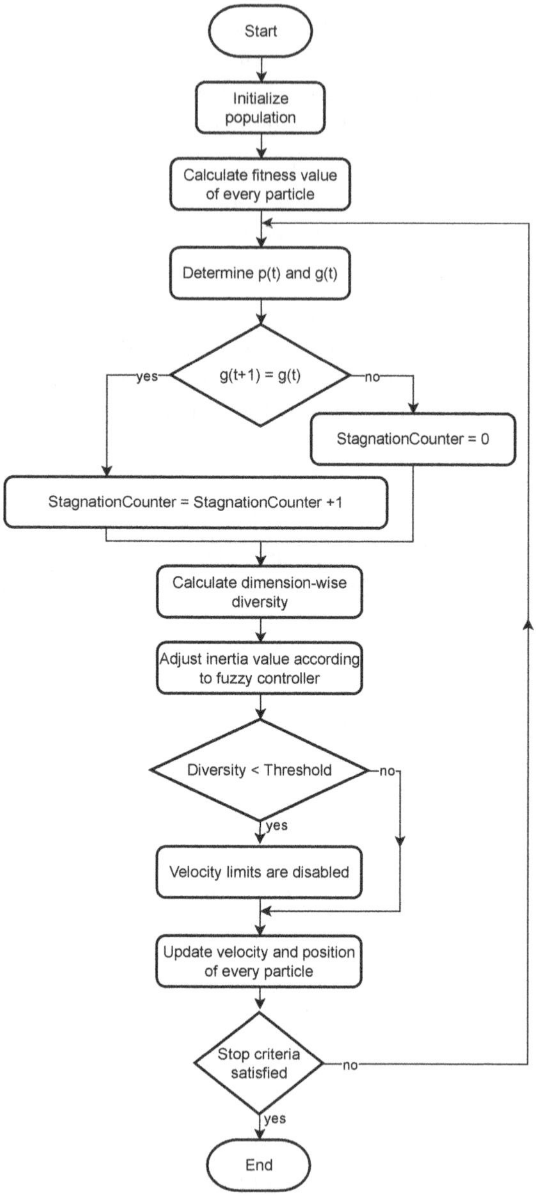

Figure 6.8: Flowchart of the proposed FCPSO.

To further optimize the exploitation phase, FCPSO introduces a diversity threshold. This threshold dynamically deactivates velocity clamping when the population diversity diminishes beyond a predefined threshold. Velocity clamping, originally introduced in PSO to mitigate velocity explosion, limits particle speeds to prevent overshooting optimal solutions. However, maintaining clamping beyond the necessary exploration phase can hinder the algorithm's ability to exploit identified solutions effectively. By deactivating clamping based on real-time diversity measurements, FCPSO ensures that particles can exploit local optima effectively when exploration goals have been achieved.

6.3.2.4 Conclusions

The proposed fuzzy-controlled PSO, FCPSO, introduces a novel approach by incorporating population diversity and a stagnation index as inputs to dynamically adjust the inertia weight parameter. Additionally, the algorithm uses population diversity to control the velocity clamping mechanism of the original PSO. These enhancements result in improved general performance, allowing FCPSO to adapt dynamically to various optimization problems.

The experiments conducted to evaluate the proposed FCPSO algorithm were designed to be comprehensive and rigorous, encompassing four distinct classes of objective functions to test various algorithmic capabilities such as exploration, exploitation, robustness, and stability. Specifically, the test suite included 13 multi-modal functions, 5 uni-modal functions, 2 composite functions, and 8 shifted functions, totaling 28 diverse and challenging functions.

Nine well-known and recent algorithms—ABC, AGPSO, CLPSO, DE, FUZZYPSO, GA, MPSO, PSO, and PSOGSA—were selected for a broad comparative evaluation. To ensure fairness, all algorithms used a population size of $N = 50$, with parameters set as per original references, optimizing performance consistency.

To statistically validate the results, each algorithm underwent 30 independent runs with a stopping criterion of 100,000 function evaluations per run. This extensive testing approach highlighted a common challenge in high-dimensional optimization problems, where many algorithms tend to underperform. By setting each optimization problem to 50 dimensions, it was possible to showcase the robustness of FCPSO in handling high-dimensional spaces effectively.

Extensive benchmarking against multimodal, unimodal, composite, and shifted functions demonstrated that FCPSO significantly outperforms not only the original PSO algorithm but also other state-of-the-art methodologies such as AGPSO, CLPSO, FUZZYPSO, MPSO, and PSOGSA. The results, particularly in high-dimensional settings, underscore the algorithm's robustness and effectiveness.

6.3.3 *Parameter Control in DE Algorithm*

Metaheuristic algorithms play a crucial role in solving complex optimization problems, especially those involving multidimensional search spaces [62]. These algorithms utilize various operators and rules to explore and exploit the search space, ultimately aiming to find optimal solutions. Among these, the Differential Evolution (DE) algorithm [30] stands out due to its effective use of crossover, mutation, and selection processes to iteratively refine candidate solutions.

DE's broad applicability spans multiple fields, including feature selection for classification [63], controller design in power electronics, and parameter extraction for photovoltaic models [64]. Despite its widespread use, DE, like other optimization algorithms, often encounters the challenge of premature convergence, which can lead to suboptimal solutions [65]. This issue is frequently attributed to a loss of diversity within the population, a critical factor for thorough exploration of the search space [66]. Maintaining diversity provides essential information that guides exploration decisions throughout the algorithm's iterations.

The Gini index (GI) [67], a statistical measure of distribution equity, ranges from 0 (perfect equality) to 1 (perfect inequality) and is represented graphically by the Lorenz curve. Although traditionally used to assess inequality in economic contexts, the Gini index has untapped potential in the realm of metaheuristic algorithms. It serves as a powerful tool for quantifying diversity within a population, offering insights into the distribution of solution qualities.

In DE, two primary parameters significantly influence its behavior: the amplification or scaling factor (F) and the crossover rate (pCR). The amplification factor (F) is pivotal during the mutation stage and has been extensively studied due to its impact on the algorithm's efficacy [68]. The crossover rate (pCR) dictates the likelihood of applying the crossover operator, which generates new solutions by combining existing ones. Despite its importance, both F and pCR are traditionally fixed throughout the algorithm's execution [69, 70].

This chapter introduces an innovative approach to DE, termed the inequality-based multi-population DE (IMDE). This approach seeks to dynamically adjust F and pCR using the dimension-wise diversity metric and the Gini index. Additionally, instead of operating on a single population, IMDE divides the main population into several subpopulations, or settlements, utilizing the K-means algorithm for clustering. The Gini index, based on fitness values, quantifies inequality, while dimension-wise diversity assesses settlement variation, providing insights into balance dynamics.

IMDE dynamically adjusts F and pCR based on subpopulation diversity and Gini index (GI). GI quantifies fitness distribution, while diversity metrics assess solution spread. Lower GI values, signaling higher equality, increase crossover likelihood, enhancing diversity and preventing premature convergence.

This dynamic adaptation of parameters based on diversity and GI aims to enhance DE's performance across various optimization problems. The proposed IMDE algorithm promises to address the limitations of fixed parameter values by fostering a more adaptive and resilient search process. By continuously tuning the F and pCR parameters in response to the evolving characteristics of subpopulations, IMDE ensures a balanced exploration and exploitation of the search space, leading to improved optimization outcomes.

6.3.3.1 Inequality-based Multi-population DE

The success of Differential Evolution (DE) depends on balancing exploration and exploitation, influenced by key parameters like scaling factor (F) and crossover rate (pCR). While traditionally constant, dynamic adjustments can enhance performance. The inequality-based multi-population DE (IMDE) introduces adaptive tuning using the Gini index (GI) and dimension-wise diversity, optimizing search efficiency.

The methodology behind IMDE is structured around three core principles:

- Reduce parameter tuning: Reducing external initialization parameters simplifies setup, improves usability, and enhances adaptability. Minimizing reliance on predefined parameters makes algorithms more accessible, especially for users without extensive tuning expertise.
- Dynamic Population Structuring: K-means clustering is used to form multiple subpopulations (settlements) during the initialization, enabling Differential Evolution (DE) to operate independently within each group.
- Adaptive Parameter Control: IMDE adjusts scaling factor (F) using dimension-wise diversity and modifies crossover rate (pCR) based on Gini index (GI) to reflect fitness inequality in subpopulations.

The general structure of the IMDE algorithm, depicted in Fig. 6.9, encompasses both the initialization and the iterative process. Each step is designed to optimize the dynamic adaptation of key parameters, ensuring robust search capabilities.

During the initialization phase, two critical parameters are set: the number of particles (N) and the number of settlements (k). Particles, normally distributed over the search space, are randomly generated and clustered into k settlements using the K-means algorithm. This clustering ensures that settlements are not uniformly sized, promoting

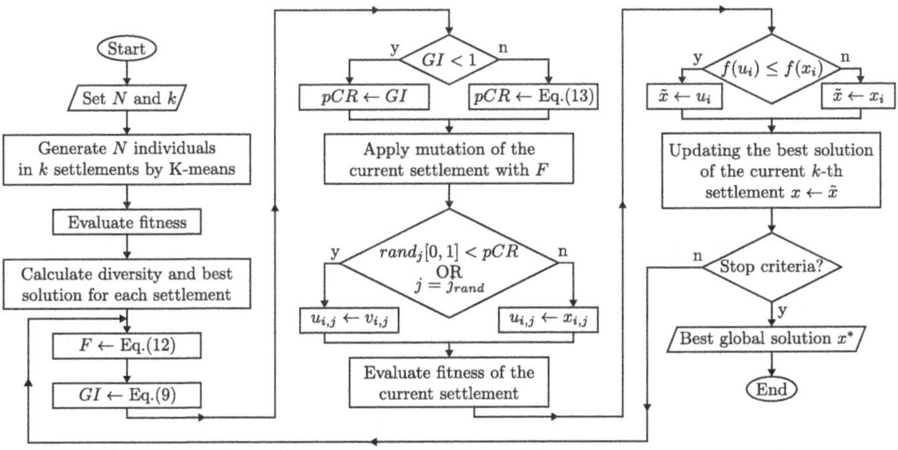

Figure 6.9: Flowchart of the proposed IMDE algorithm.

diversity. A sensitivity analysis determines the optimal number of settlements. The final step involves calculating the diversity and identifying the best solution within each settlement.

On the other hand, during the iterative process, the scaling factor (F) is dynamically calculated for each settlement using dimension-wise diversity. Subsequently, the GI is computed and utilized dynamically as the crossover rate (pCR) value. The standard DE stages are then performed independently for each settlement. This process continues until the stopping criteria are met, and the best solution is selected.

6.3.3.2 Evolution of the Scaling Factor Using Dimension-Wise Diversity

As previously explained, the scaling factor F is typically set to fixed values such as 0.8 or 0.9. However, fixed values can lead to premature convergence. In IMDE, dimension-wise diversity is employed to dynamically adjust F. This approach measures the diversity among elements in each settlement during each iteration, normalizing it between 0 and 1. The scaling factor F is then calculated as:

$$F = div_{min} + (div_{max} - div_{min}) \cdot rand \tag{6.1}$$

where, $rand$ is a random number between 0 and 1, allowing F to vary within the range of div_{min} and div_{max}. Figure 6.10 illustrates how F evolves for different functions.

In the initial stages, high diversity results in higher values of F, facilitating exploration. As diversity decreases, F values spread across a broader range, balancing exploration and exploitation. Finally, when diversity is very low, F values fluctuate significantly, enabling the algorithm to refine solutions while maintaining some exploratory capability.

The crossover probability (pCR) is another critical parameter. Traditionally set to low values like 0.2 or 0.3, IMDE adjusts pCR dynamically using the Ginny index. The GI, normalized between 0 and 1, measures the inequality of fitness values within each settlement.

Two scenarios govern the calculation of pCR:

Case 1: If the GI is less than or equal to 1, pCR is set equal to the GI. Case 2: If the GI cannot be calculated, pCR is chosen randomly within the range [0.1, 0.5]:

$$pCR = 0.1 + (0.5 - 0.1) \cdot rand \tag{6.2}$$

Equation 6.2 demonstrates the dynamic adjustment of pCR based on the GI. In the early stages, low GI values indicate high diversity, resulting in lower pCR values, which is beneficial for exploration. As the algorithm progresses and diversity decreases, pCR values increase, enhancing the probability of crossover and promoting exploitation of promising areas.

By dynamically adjusting F and pCR based on diversity and the GI, IMDE ensures a robust balance between exploration and exploitation throughout the optimization process. This adaptive approach significantly improves the DE algorithm's ability to find optimal solutions in complex, multidimensional search spaces.

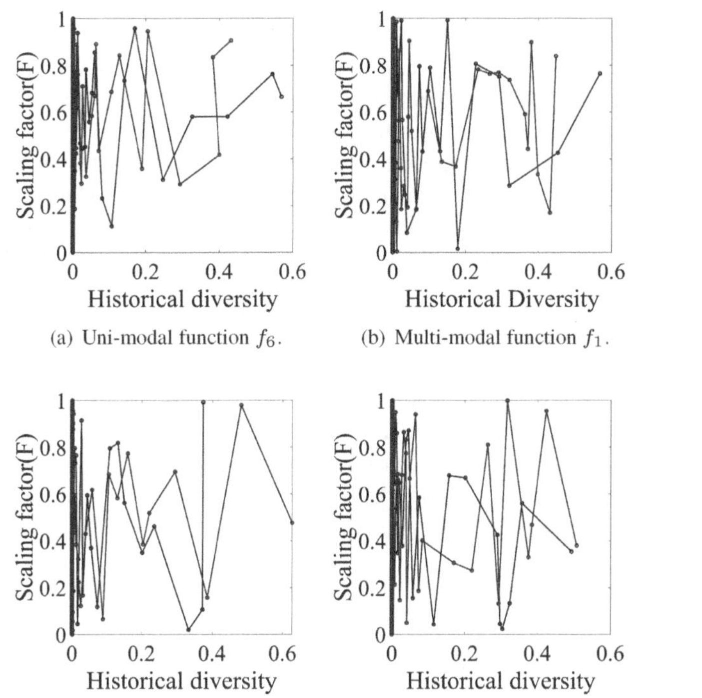

Figure 6.10: Dynamic adjustment of F based on the diversity of each settlement.

6.3.3.3 Conclusions

In this section, an enhanced Differential Evolution (DE) algorithm, termed IMDE, was introduced. It features a multi-population strategy and a dynamic adjustment mechanism for the parameters pCR and F, which are crucial for crossover and mutation operations. The multi-population framework leverages K-means clustering to create diverse subpopulations or settlements, each characterized by varying fitness levels. The Gini index, a measure of statistical dispersion representing the inequality of fitness values within a settlement, is computed for each subpopulation. These diversity metrics and the Gini index inform the adaptive tuning of pCR and F, allowing the algorithm to respond dynamically to the search landscape.

The scaling factor F influences the amplitude of changes introduced during the mutation process. By adjusting F based on the diversity of subpopulations, IMDE maintains a balance between exploration and exploitation, thus enhancing its adaptability to different optimization challenges. The crossover probability pCR, influenced by the Gini index, ensures that the algorithm sustains sufficient diversity throughout the search process, preventing premature convergence and enhancing solution robustness.

An extensive experimental evaluation involved 43 benchmark functions, categorized into unimodal, multimodal, composite, and shifted functions, to comprehensively assess the algorithm's performance. Comparative analyses with several state-of-the-art DE and Particle Swarm Optimization (PSO) variants, including JADE, L-SHADE, EDA, AGPSO, IPSO, MPSO, and standard DE and PSO, demonstrated the superior performance of IMDE. The proposed algorithm achieved the best results in a significant majority of the test functions, particularly excelling in unimodal and multimodal functions with 30 dimensions. It also showed competitive performance in composite and shifted functions, consistently achieving optimal or near-optimal solutions.

The dynamic adjustment of parameters based on population diversity and the Gini index proved to be a robust strategy, enabling IMDE to efficiently navigate complex search spaces and avoid common pitfalls such as local optima and premature convergence. The results underscore the efficacy of integrating adaptive mechanisms within the DE framework, highlighting the potential of IMDE for a wide range of optimization problems. Future work could explore further enhancements, such as integrating additional diversity measures or hybridizing IMDE with other optimization techniques, to further bolster its performance and applicability.

References

[1] S. S. Rao, *Engineering Optimization: Theory and Practice.* Wiley, 2019.

[2] X.-S. Yang, S. Deb, and S. Fong, "Metaheuristic algorithms: Optimal balance of intensification and diversification," *Applied Mathematics & Information Sciences*, vol. 7, no. 5, pp. 1697–1703, 2013.

[3] U. Farooq, A. Shahzad, and M. A. Mirza, "Metaheuristic techniques in image processing: a review," *Artificial Intelligence Review*, vol. 42, no. 4, pp. 1–50, 2014.

[4] A. K. Bhandari, A. Kumar, and G. K. Singh, "A novel image multithresholding based segmentation using nature inspired optimization algorithms," *Expert Systems with Applications*, vol. 63, pp. 112–133, 2015.

[5] A. E. Eiben and J. E. Smith, *Introduction to Evolutionary Computing.* Springer, 2015.

[6] H. Peng and X. Xiao, "Applications of metaheuristic optimization algorithms in image processing," *Expert Systems with Applications*, vol. 41, no. 6, pp. 3005–3021, 2014.

[7] R. A. Alshammari and K. M. Elleithy, "Optimizing wireless network protocols using genetic algorithms and other metaheuristic techniques: A survey," *Wireless Personal Communications*, vol. 77, no. 3, pp. 1759–1786, 2014.

[8] T. G. Crainic, G. Perboli, and R. Tadei, "Metaheuristics for a class of demand-management transportation problems," *Transportation Science*, vol. 43, no. 3, pp. 291–302, 2009.

[9] I. Fister, I. Fister Jr, X.-S. Yang, and J. Brest, "A comprehensive review of cuckoo search: variants and hybrids," *Applied Soft Computing*, vol. 11, no. 3, pp. 1972–1983, 2013.

[10] T. Chen and C. Guestrin, "Xgboost: A scalable tree boosting system," in *Proceedings of the 22nd ACM SIGKDD International Conference on Knowledge Discovery and Data Mining*. ACM, 2016, pp. 785–794.

[11] E.-G. Talbi, *Metaheuristics: From Design to Implementation*. Wiley, 2009.

[12] Z. Pan, J. Polden, N. Larkin, S. Van Duin, and J. Norrish, "Recent progress on programming methods for industrial robots," *Robotics and Computer-Integrated Manufacturing*, vol. 28, no. 2, pp. 87–94, 2012.

[13] F. L. Oliveira, E. C. da Silva, and J. P. Papa, "A survey on intelligent systems for medical diagnosis," *Artificial Intelligence Review*, vol. 41, no. 1, pp. 89–113, 2014.

[14] E. Z. de Medeiros, M. J. B. de Almeida, and F. B. do Nascimento, "Metaheuristic approaches to the diagnosis of diseases," *Computer Methods and Programs in Biomedicine*, vol. 126, pp. 59–71, 2016.

[15] K. Hussain, M. N. M. Salleh, S. Cheng, and Y. Shi, "Metaheuristic research: a comprehensive survey," *Artificial Intelligence Review*, vol. 52, pp. 2191 – 2233, 2018. [Online]. Available: https://api.semanticscholar.org/CorpusID:32526909

[16] K. Sörensen, "Metaheuristics—the metaphor exposed," *International Transactions in Operational Research*, vol. 22, no. 1, pp. 3–18, 2015. [Online]. Available: https://onlinelibrary.wiley.com/doi/abs/10.1111/itor.12001

[17] J. Xu and J. Zhang, "Exploration-exploitation tradeoffs in metaheuristics: Survey and analysis," in *Proceedings of the 33rd Chinese Control Conference (CCC)*. IEEE, 2014, pp. 8633–8638.

[18] M. Črepinšek, S.-H. Liu, and M. Mernik, "Exploration and exploitation in evolutionary algorithms," *ACM Computing Surveys (CSUR)*, vol. 45, no. 3, pp. 1–33, Jun 2013.

[19] S.-R. Chen, A. Boluf, J. Montgomery, and T. Hendtlass, "An analysis on the effect of selection on exploration in particle swarm optimization and differential evolution," in *2019 IEEE Congress on Evolutionary Computation (CEC)*. IEEE, 2019, pp. 3037–3044.

[20] F. Fausto, A. Reyna-Orta, E. Cuevas, G. García-Álvarez, and M. Perez-Cisneros, *From Ants to Whales: Metaheuristics for all Tastes*. Springer Netherlands, 2019.

[21] X.-S. Yang, *Swarm-Based Metaheuristic Algorithms and No-Free-Lunch Theorems*. IntechOpen, 2012.

[22] ——, "Nature-inspired metaheuristic algorithms: Success and new challenges," *Journal of Computational Engineering & Information Technology*, vol. 1, no. 1, 2012.

[23] H.-P. Kriegel, E. Schubert, and A. Zimek, "The (black) art of runtime evaluation: Are we comparing algorithms or implementations?" *Knowledge and Information Systems*, vol. 52, no. 2, pp. 341–378, 2017.

[24] S. Cheng, Y. Shi, Q. Qin, Q. Zhang, and R. Bai, "Population diversity maintenance in brain," *Brain*, vol. 4, no. 2, pp. 83–97, 2015.

[25] T. Al-Quraishi, J. H. Abawajy, M. U. Chowdhury, S. Rajasegarar, and A. S. Abdalrada, *Recent Advances on Soft Computing and Data Mining*, ser. Advances in Intelligent Systems and Computing. Springer International Publishing, 2018, vol. 700.

[26] M. N. M. Salleh *et al.*, "Exploration and exploitation measurement in swarm-based metaheuristic algorithms: An empirical analysis," in *Advances in Intelligent Systems and Computing*. Springer, 2018, vol. 700, no. Scdm, pp. 24–32.

[27] X. S. Yang, S. Deb, and S. Fong, "Metaheuristic algorithms: Optimal balance of intensification and diversification," *Applied Mathematics & Information Sciences*, vol. 8, no. 3, pp. 977–983, 2014.

[28] X. S. Yang, S. Deb, T. Hanne, and X. He, "Attraction and diffusion in nature-inspired optimization algorithms," *Neural Computing and Applications*, vol. 19, 2015.

[29] H. Du, Z. Wang, and W. E. I. Zhan, "Elitism and distance strategy for selection of evolutionary algorithms," *IEEE Access*, vol. 6, pp. 44 531–44 541, 2018.

[30] R. Storn and K. Price, "Differential evolution – a simple and efficient heuristic for global optimization over continuous spaces," *Journal of Global Optimization*, vol. 11, no. 4, pp. 341–359, 1997.

[31] T. Huang, Z.-H. Zhan, X. Jia, H. Yuan, J. Jiang, and J. Zhang, "Niching community based differential evolution for multimodal optimization problems," in *2017 IEEE Symposium Series on Computational Intelligence (SSCI)*, 2017, pp. 1–8.

[32] R. Poli, J. Kennedy, and T. Blackwell, "Particle swarm optimization," *Swarm Intelligence*, vol. 1, no. 1, pp. 33–57, Oct 2007.

[33] S. Mirjalili, S. M. Mirjalili, and A. Lewis, "Grey wolf optimizer," *Advances in Engineering Software*, vol. 69, pp. 46–61, Mar 2014.

[34] X.-S. Yang, "Firefly algorithms for multimodal optimization," in *Lecture Notes in Computer Science (including subseries Lecture Notes in Artificial Intelligence and Lecture Notes in Bioinformatics)*, vol. 5792, 2009, pp. 169–178.

[35] E. Rashedi, H. Nezamabadi-pour, and S. Saryazdi, "Gsa: A gravitational search algorithm," *Information Sciences*, vol. 179, no. 13, pp. 2232–2248, 2009.

[36] D. Karaboga, "An idea based on honey bee swarm for numerical optimization," Erciyes University, Tech. Rep. TR06, 2005, technical Report.

[37] D. Karaboga and B. Basturk, "A powerful and efficient algorithm for numerical function optimization: artificial bee colony (abc) algorithm," *Journal of global optimization*, vol. 39, no. 3, pp. 459–471, 2007.

[38] X.-S. Yang, "A new metaheuristic bat-inspired algorithm," in *Nature Inspired Cooperative Strategies for Optimization (NICSO 2010)*, 2010, pp. 65–74.

[39] N. Hansen and A. Ostermeier, "Completely derandomized self-adaptation in evolution strategies," *Evolutionary Computation*, vol. 9, no. 2, pp. 159–195, Jun 2001.

[40] A. Askarzadeh, "A novel metaheuristic method for solving constrained engineering optimization problems: Crow search algorithm," *Computers & Structures*, vol. 169, pp. 1–12, 2016.

[41] E. Cuevas, M. Cienfuegos, D. Zaldívar, and M. Pérez-Cisneros, "A swarm optimization algorithm inspired in the behavior of the social-spider," *Expert Systems with Applications*, vol. 40, no. 16, pp. 6374–6384, 2013.

[42] R. V. Rao, V. J. Savsani, and D. P. Vakharia, "Teaching-learning-based optimization: An optimization method for continuous non-linear large scale problems," *Information Sciences*, vol. 183, no. 1, pp. 1–15, 2012.

[43] S. Mirjalili, "Moth-flame optimization algorithm: A novel nature-inspired heuristic paradigm," *Knowledge-Based Systems*, vol. 89, pp. 228–249, 2015.

[44] B. Chopard and M. Tomassini, "Particle swarm optimization," in *Natural Computing Series*, 2018, pp. 97–102.

[45] S. Mirjalili and A. Lewis, "The whale optimization algorithm," *Advances in Engineering Software*, vol. 95, pp. 51–67, 2016.

[46] J. Derrac, S. García, D. Molina, and F. Herrera, "A practical tutorial on the use of nonparametric statistical tests as a methodology for comparing evolutionary and swarm intelligence algorithms," *Swarm and Evolutionary Computation*, vol. 1, no. 1, pp. 3–18, 2011.

[47] N. Hansen, "The cma evolution strategy: A tutorial," *arXiv preprint arXiv:1604.00772*, pp. 75–102, 2016.

[48] S. Das and P. N. Suganthan, "Differential evolution: A survey of the state-of-the-art," *IEEE Transactions on Evolutionary Computation*, vol. 15, no. 1, pp. 4–31, 2011.

[49] I. Fister, X. S. Yang, and J. Brest, "A comprehensive review of firefly algorithms," *Swarm and Evolutionary Computation*, vol. 13, pp. 34–46, 2013.

[50] F. Marini and B. Walczak, "Particle swarm optimization (pso). a tutorial," *Chemometrics and Intelligent Laboratory Systems*, vol. 149, pp. 153–165, 2015. [Online]. Available: https://www.sciencedirect.com/science/article/pii/S0169743915002117

[51] E. Cuevas, M. Cienfuegos, D. Zaldívar, and M. Pérez-Cisneros, "A swarm optimization algorithm inspired in the behavior of the social-spider," *Expert Systems with Applications*, vol. 40, no. 16, pp. 6374–6384, 2013. [Online]. Available: https://www.sciencedirect.com/science/article/pii/S0957417413003394

[52] Y. Shi and R. C. Eberhart, "Fuzzy adaptive particle swarm optimization," in *Proceedings of the 2001 congress on evolutionary computation (IEEE Cat. No. 01TH8546)*, vol. 1. IEEE, 2001, pp. 101–106.

[53] P. Melin, F. Olivas, O. Castillo, F. Valdez, J. Soria, and M. Valdez, "Optimal design of fuzzy classification systems using pso with dynamic parameter adaptation through fuzzy logic," *Expert Systems with Applications*, vol. 40, no. 8, pp. 3196–3206, 2013.

[54] D.-p. Tian and N.-q. Li, "Fuzzy particle swarm optimization algorithm," in *2009 International Joint Conference on Artificial Intelligence*. IEEE, 2009, pp. 263–267.

[55] S. Cheng, Y. Shi, Q. Qin, Q. Zhang, and R. Bai, "Population diversity maintenance in brain storm optimization algorithm," *Journal of Artificial Intelligence and Soft Computing Research*, vol. 4, no. 2, pp. 83–97, 2014.

[56] A. Aleti and I. Moser, "A systematic literature review of adaptive parameter control methods for evolutionary algorithms," *ACM Computing Surveys (CSUR)*, vol. 49, no. 3, pp. 1–35, 2016.

[57] T. M. Shami, A. A. El-Saleh, M. Alswaitti, Q. Al-Tashi, M. A. Summakieh, and S. Mirjalili, "Particle swarm optimization: A comprehensive survey," *IEEE Access*, vol. 10, pp. 10 031–10 061, 2022.

[58] A. Saffari, M. Khishe, and S. H. Zahiri, "Fuzzy-choa: an improved chimp optimization algorithm for marine mammal classification using artificial neural network," *Analog Integrated Circuits and Signal Processing*, vol. 111, pp. 403–417, 6 2022. [Online]. Available: https://link.springer.com/article/10.1007/s10470-022-02014-1

[59] B. Morales-Castañeda, D. Zaldívar, E. Cuevas, F. Fausto, and A. Rodríguez, "A better balance in metaheuristic algorithms: Does it exist?" *Swarm and Evolutionary Computation*, vol. 54, p. 100671, 2020. [Online]. Available: https://www.sciencedirect.com/science/article/pii/S2210650219304080

[60] H. Jia, K. Sun, W. Zhang, and X. Leng, "An enhanced chimp optimization algorithm for continuous optimization domains," *Intelligent Systems*, vol. 8, no. 1, p. 65–82, 2021.

[61] L. Li, Y. Yu, S. Bai, J. Cheng, and X. Chen, "Towards effective network intrusion detection: A hybrid model integrating gini index and gbdt with pso," *Journal of Sensors*, vol. 2018, p. 1–9, 2018.

[62] Z.-H. Zhan, L. Shi, K. C. Tan, and J. Zhang, “A survey on evolutionary computation for complex continuous optimization,” *Artificial Intelligence Review*, vol. 55, no. 1, pp. 59–110, 2022.

[63] P. Wang, B. Xue, J. Liang, and M. Zhang, “Differential evolution with duplication analysis for feature selection in classification,” *IEEE Transactions on Cybernetics*, 2022.

[64] Z. Hu, W. Gong, and S. Li, “Reinforcement learning-based differential evolution for parameters extraction of photovoltaic models,” *Energy Reports*, vol. 7, pp. 916–928, 2021.

[65] B. Morales-Castaneda, O. Maciel-Castillo, M. A. Navarro, I. Aranguren, A. Valdivia, A. Ramos-Michel, D. Oliva, and S. Hinojosa, “Handling stagnation through diversity analysis: A new set of operators for evolutionary algorithms,” in *2022 IEEE Congress on Evolutionary Computation (CEC)*. IEEE, 2022, pp. 1–7.

[66] D. Oliva, E. Rodriguez-Esparza, M. S. Martins, M. Abd Elaziz, S. Hinojosa, A. A. Ewees, and S. Lu, “Balancing the influence of evolutionary operators for global optimization,” in *2020 IEEE Congress on Evolutionary Computation (CEC)*. IEEE, 2020, pp. 1–8.

[67] F. A. Farris, “The gini index and measures of inequality,” *The American Mathematical Monthly*, vol. 117, no. 10, pp. 851–864, 2010.

[68] K. Opara and J. Arabas, “Comparison of mutation strategies in differential evolution–a probabilistic perspective,” *Swarm and Evolutionary Computation*, vol. 39, pp. 53–69, 2018.

[69] D. Zaharie, “Influence of crossover on the behavior of differential evolution algorithms,” *Applied Soft Computing*, vol. 9, no. 3, pp. 1126–1138, 2009.

[70] D. Bajer, “Parameter control for differential evolution by storage of successful values at an individual level,” *Journal of Computational Science*, vol. 68, p. 101985, 2023.

Chapter 7

Different Indicators for Measuring Diversity

7.1 Introduction

From the previous chapter, it can be concluded that maintaining population diversity in evolutionary computation approaches poses a significant challenge throughout optimization. A balanced exploration-exploitation dynamic is crucial for achieving optimal results, and a diverse population proportion plays a key role in achieving this balance. Moreover, implementing processes to maintain diversity is key to solving complex problems and ensuring robust and competitive solutions.

Various methods assess population diversity throughout optimization. [1] introduces a surrogate hypervolume-based metric for selecting candidate solutions. [2, 3] apply Dimension-Wise Diversity (DWD) to the Artificial Bee Colony (ABC) [4] algorithm. [5] refines PSO using a diversity-guided multilevel learning strategy, incorporating chaotic opposition-based particle distribution. [6, 7] enhance DE through Average Population Diversity (APD), dynamically adjusting population size. [8] presents an auto-enhanced diversity technique, detecting premature convergence and stagnation in DE.

Literature highlights distance between all possible chromosome pairs as a core diversity metric [9]. If all chromosomes are identical, diversity is absent, while greater variation increases diversity. Population diversity is defined by differences among individuals, guiding diversity mechanisms through:

- Promoting variation within the population.
- Exhaustive comparison of individual differences.

DOI: 10.1201/9781032723563-7

Focus on differences: Diversity mechanisms seek to promote the existence of differences among individuals in a population.

Exhaustive comparison: Diversity assessment involves analyzing the differences between every possible pair of individuals.

Diversity mechanisms fall into three main types:

- **Distance-based methods:** Measure solution dispersion, indicating exploration levels.
- **Entropy-based methods:** Evaluate uniformity of distribution across the search space.
- **Variance-based methods:** Analyze statistical variation in decision variables.

Diversity methods—distance, entropy, and variance-based—can be applied to entire populations or subpopulations at different optimization stages. These techniques often overlap, enhancing diversity assessment to fit specific problem needs. Upcoming sections delve into each category, detailing key metrics for evaluating diversity in optimization.

7.2 Distance-based Methods

Various diversity metrics have been proposed to assess and maintain population diversity, each with advantages and limitations. One of the most widely used diversity measures is based on distance metrics, which quantify how far apart solutions are in the solution space. This section will explore three popular distance-based diversity measures: Euclidean distance, Hamming distance, and Average population diversity.

7.2.1 *Euclidean Distance*

The Euclidean Distance is a widely used metric in computer vision, machine learning, and data analysis [10]. Based on the Pythagorean theorem, it calculates the shortest path between two points in n-dimensional space. See Fig. 7.1 for a visual representation.

Figure 7.1 illustrates the calculation of the 2D Euclidean distance between two points, point A with coordinates (1, 1) and point B with coordinates (4, 4). A dashed line represents the shortest path between them (Euclidean distance), while the black straight lines represent the Pythagorean theorem in action. In the upper part of the figure, it can be seen that the result of the distance between these two points is 4.24. This is easily calculated using the following Euclidean formula for 2-dimensions:

$$d = \sqrt{(x_2 - x_1)^2 + (y_2 - y_1)^2} \tag{7.1}$$

where, d is the distance, x_1 and x_2 are the X-axis coordinates and y_1 and y_2 the Y-axis coordinates. In metaheuristic algorithms, by calculating the Euclidean distances between all pairs of solutions, we can obtain a measure of the overall dispersion

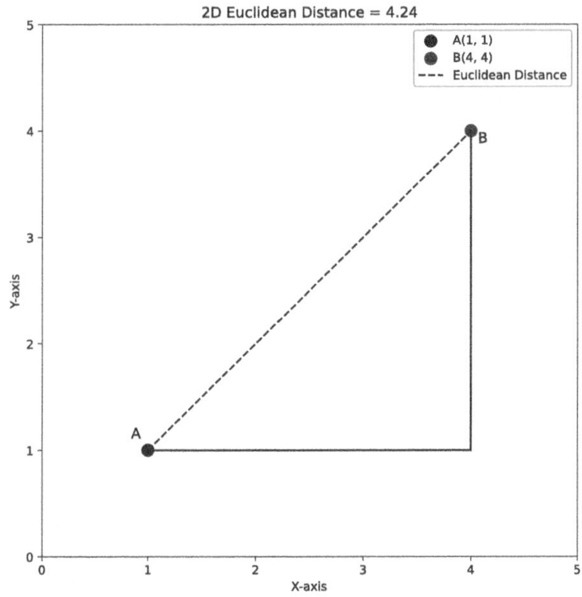

Figure 7.1: Representation of the Euclidean distance between two points.

of the population. However, search spaces are usually more complicated than two dimensions. The following equation is used to calculate the distances between points in an n-dimensional space:

$$d(\mathbf{a}, \mathbf{b}) = \sqrt{\sum_{i=1}^{n} (a_i - b_i)^2} \tag{7.2}$$

where, $\mathbf{a} = (a_1, a_2, ..., a_n)$ and $\mathbf{b} = (b_1, b_2, ..., b_n)$ are the points into which it assesses the distance. $a_i - b_i$ is the difference between the coordinates of the two points in the dimension i, and n defines the number of coordinates needed to describe the position of the points. Equation 7.2 must be adapted to quantify the total diversity of the points within a multidimensional space as follows:

$$d_{\text{ET}} = \sum_{i=1}^{N} \sum_{j=i+1}^{N} \sqrt{\sum_{k=1}^{D} (A_{i,k} - A_{j,k})^2} \tag{7.3}$$

From Equation 7.3, the total diversity d_{ET} is calculated as the sum of the Euclidean distances between all set pairs S. Where D is the number of dimensions of the space, N is the total number of points in the set S, $A_i = (A_{i1}, A_{i2}, ..., A_{in})$ and $A_j = (A_{j1}, A_{j2}, ..., A_{jn})$ are the coordinates of points A_i and A_j in the multidimensional space, i is the index of the first point A_i and j is the index of the second point A_j. This equation can be adapted to calculate the average distances between all pairs of points instead of the total sum of the distances (total diversity) by dividing the total diversity d_{ET} by the number of unique pairs of points.

A practical example for calculating total and average diversity based on Euclidean distance is presented to understand this concept better. Assume that five solution points are distributed in the following positions in a three-dimensional space:

$$A_1 = (1, 2, 3)$$

$$A_2 = (0, 1, 9)$$

$$A_3 = (6, 8, 2)$$

$$A_4 = (7, 1, 4)$$

$$A_5 = (9, 7, 1)$$

The graphical three-dimensional representation for these solutions is depicted in Fig. 7.2. The first step is to compute the distance between all the pairs of solutions, as depicted in Equation 7.3. Since there exist five solution points, the total unique pairs would be $5(5-1)/2 = 10$. The calculation for each unique pair is presented as follows:

$$d(A_1, A_2) = \sqrt{(1-0)^2 + (2-1)^2 + (3-9)^2} = \sqrt{38} \approx 6.16$$

$$d(A_1, A_3) = \sqrt{(1-6)^2 + (2-8)^2 + (3-2)^2} = \sqrt{62} \approx 7.87$$

$$d(A_1, A_4) = \sqrt{(1-7)^2 + (2-1)^2 + (3-4)^2} = \sqrt{38} \approx 6.16$$

$$d(A_1, A_5) = \sqrt{(1-9)^2 + (2-7)^2 + (3-1)^2} = \sqrt{93} \approx 9.64$$

$$d(A_2, A_3) = \sqrt{(0-6)^2 + (1-8)^2 + (9-2)^2} = \sqrt{134} \approx 11.58$$

$$d(A_2, A_4) = \sqrt{(0-7)^2 + (1-1)^2 + (9-4)^2} = \sqrt{74} \approx 8.60$$

$$d(A_2, A_5) = \sqrt{(0-9)^2 + (1-7)^2 + (9-1)^2} = \sqrt{181} \approx 13.45$$

$$d(A_3, A_4) = \sqrt{(6-7)^2 + (8-1)^2 + (2-4)^2} = \sqrt{54} \approx 7.34$$

$$d(A_3, A_5) = \sqrt{(6-9)^2 + (8-7)^2 + (2-1)^2} = \sqrt{11} \approx 3.31$$

$$d(A_4, A_5) = \sqrt{(7-9)^2 + (1-7)^2 + (4-1)^2} = \sqrt{49} = 7$$

After obtaining the distance results for all pairs, they are summed to calculate the total diversity:

$$d_{ET} = 6.16 + 7.87 + 6.16 + 9.64 + 11.58 + 8.60 + 13.45 + 7.34 + 3.31 + 7$$

$$d_{ET} = 81.11$$

Finally, the average diversity for all pairs is calculated as $d_{avg} = \frac{d_{ET}}{10} = 8.11$.

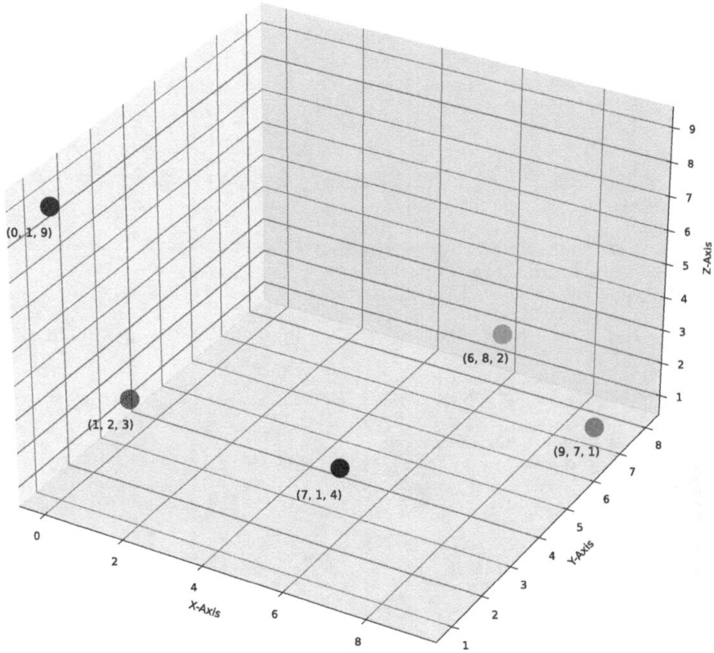

Figure 7.2: Representation of solution points in 3-dimensional space.

As a result, a higher average Euclidean distance indicates a more diverse population, which can also depict an exploration phase of the search space. In turn, a lower average distance suggests a more homogeneous population, potentially indicating premature convergence towards a local optimum. Nonetheless, relying on average distance alone can be misleading. While clusters of solutions may be similar to each other, they can be far apart from other clusters. Therefore, depending on the application, an analysis of the distribution of Euclidean distances, rather than just the average, can provide a more subtle understanding of population diversity.

In conclusion, the Euclidean distance is a simple and intuitive metric that can give an insight into the diversity of the population throughout the optimization process, although it does have some limitations. One key limitation is that it is sensitive to the scale of the variables in the solution space. This issue can be addressed by normalizing the variables and ensuring that all elements contribute equally to the diversity measure. In turn, Euclidean distance only calculates spatial separation and does not account for the quality of the solutions based on the objective function. Thus, two solutions far apart in the solution space may still be suboptimal. Despite its limitations, Euclidean distance's simplicity, versatility, and intuitive nature make it a widely used tool for assessing and maintaining diversity in optimization tasks. Subsections 7.2.1.1 and 7.2.1.2 provide pseudocode and an example implementation in MATLAB® to enhance the understanding of the concepts discussed.

7.2.1.1 Pseudocode

```
BEGIN Calculate Diversity based on Euclidean distance

    # Define the problem
    Define the population size N
    Define the dimensionality of the search space D

    # Initialize variables
    sum_distances = 0      # Accumulator for the sum of distances
    pair_count = 0         # Counter for the number of unique pairs

    # Calculate pairwise Euclidean distances
    FOR each individual i in Population P (i = 1 to N)
        FOR each individual j in Population P (j = i+1 to N)
            # Compute Euclidean distance
            distance = SQRT(SUM((P[i][d] - P[j][d])^2 FOR d = 1 to D))

            # Accumulate the distance
            sum_distances = sum_distances + distance #Total diversity

            # Increment the pair count
            pair_count = pair_count + 1
        END FOR
    END FOR

    # Compute average diversity
    average_diversity = sum_distances / pair_count

    # Output the result
    RETURN average_diversity

END Calculate Diversity based on Euclidean distance
```

Pseudocode explanation

- Input Parameters: The function defines the number of solutions (N), the number of decision variables (D). The variables sum_distances and pair_count are initialized to sum the distances of the solutions and count the total pairs of solutions, respectively.
- Calculating the diversity: A nested pair of loops runs n times over each solution pair. For each iteration, the Euclidean distance is calculated, and the variables sum_distances and pair_count are updated to the actual value.
- Result: Once all n solutions pairs have been analyzed through the Euclidean distance diversity method, the average_diversity value is assessed by dividing the sum_distances with the pair_count. Finally, the average diversity result based on Euclidean distance calculation is returned.

7.2.1.2 MATLAB Code

```
function avgDiversity = EuclideanDistance(population)
    % EuclideanDistance = Calculates the average diversity of a population
    % based on Euclidean distance.
    % population - Matrix (NxD) where N is the number of particles and
    %              D is the dimensionality of space.

    % Population Size
    numParticles = size(population, 1);
```

```
    % Initialize the sum of distances and pair counter
    totalDistance = 0;
    pairCount = 0;

    % Calculate the Euclidean distances between each pair of particles
    for i = 1:numParticles
        for j = i+1:numParticles
            % Calculate Euclidean distance between particle i and j
            distance = sqrt(sum((population(i, :) - population(j, :)).^2));
            totalDistance = totalDistance + distance; % Sum distances
            pairCount = pairCount + 1;               % Pair Counting
        end
    end

    % Calculate average diversity
    avgDiversity = totalDistance / pairCount;

    % Display results
    fprintf('The average diversity based on the Euclidean distance is: %.4f\n',
    avgDiversity);
end
```

Example input for testing the code in MATLAB

```
% Define a population matrix of 10 individuals in a 3-dimensional space
population = [
    2, 4, 6;
    5, 1, 3;
    7, 8, 2;
    4, 9, 5;
    1, 6, 7;
    8, 3, 9;
    0, 2, 4;
    6, 7, 1;
    3, 5, 8;
    9, 0, 2
];

% Call the function
avgDiversity = EuclideanDistance(population);
```

7.2.2 *Hamming Distance*

The Hamming distance is a valuable measure in information theory. It has multiple applications in metaheuristic algorithms, such as diversity assessment, parent selection in genetic algorithms, and the design of mutation operators [11]. This distance measure counts the number of positions by which two binary vectors of the same size differ. In this context, it is a simple and effective measure that provides insight into the similarity between binary solutions. The Hamming distance d_H of two strings A and B of equal length n is calculated as follows:

$$d_H = \sum_{i=1}^{n} \mathbb{I}\left(A_i \neq B_i\right) \tag{7.4}$$

Here, A_i and B_i represent the values of position i of the strings A and B, moreover $\mathbb{I}\left(A_i \neq B_i\right)$ is a function that evaluates a conditional if the element in position i of A is is not the same as the position i of B then the resultant value is 1. In turn, if the element in position i of A is equal to the one in position i of B then the resultant is 0. The sum of the resultants would indicate how distant the two strings are. It is important to emphasize that the first step in using Hamming distance is to ensure that solutions are represented as equal-length bit strings. For example, consider a solution vector represented as $S = [42, 56, 44, 10, 58]$, then their respective binary values would be the following:

The next step is to calculate the distance between unique pairs in the population; similar to the example of the Euclidean distance, we have five solutions, so the total number of unique pairs will be $5(5-1)/2 = 10$. Figure 7.3 visually explains how the difference between points is assessed. Here, the first two binary solutions S_1 and S_2 of Table 7.1 are taken for representation. In grey color are depicted the differences among S_1 and S_2 where the Hamming distance d_H result is 2. All the possible combinations of unique pairs of the population are calculated and summarized in Table 7.2.

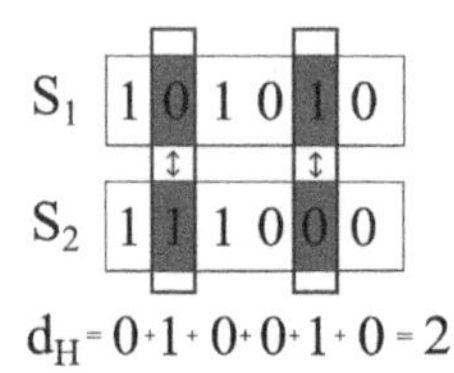

Figure 7.3: Representation of the Hamming distance calculation between two points.

Table 7.1: Binary values of solution vector S.

Solution	Binary Value	Decimal Value
S_1	101010	42
S_2	111000	56
S_3	101100	44
S_4	001010	10
S_5	111010	58

From Table 7.2 it can be observed that the total diversity based on the Hamming distance is equal to 20. The calculated value is averaged to determine the overall diversity of the solutions. This is achieved by dividing the total diversity value by the number of unique pairs within the population $\frac{20}{10} = 2$, which results in an average diversity score of 2. This result indicates that the solutions in the population vary by 2 positions. If the diversity is low (close to 0), it shows that the population is highly concentrated in one area of the search space, which could lead to premature convergence. If the diversity is high, it is then intuited that the solutions are diverse enough, promoting the exploration of the search space. In addition to the example, the Subsections 7.2.2.1 and 7.2.2.2 present the application of the Hamming distance as a diversity measure in pseudocode and MATLAB code, correspondingly.

Table 7.2: Hamming distance calculation for each pair of solutions.

Pairs	d_H
(S_1, S_2)	2
(S_1, S_3)	2
(S_1, S_4)	1
(S_1, S_5)	1
(S_2, S_3)	2
(S_2, S_4)	3
(S_2, S_5)	1
(S_3, S_4)	3
(S_3, S_5)	3
(S_4, S_5)	2
Total Diversity	**20**

The Hamming distance has been recognized as a valuable criterion for measuring the variation among candidate solutions in the evolutionary computation field. As higher Hamming distances indicate, a diverse population has been shown to encourage broader solution space exploration and to help avoid premature convergence to less optimal solutions. However, it should be noted that the Hamming distance focuses on pairwise differences and does not capture the spatial distribution or structural diversity of solutions. Therefore, combining the Hamming distance with other metrics, such as the Euclidean distance or Shannon entropy, can offer a better understanding of population diversity.

7.2.2.1 Pseudocode

```
BEGIN Calculate Diversity based on the Hamming Distance

    # Define the problem
    Define the population size N
    Define the length of each solution L (binary representation length)

    # Initialize variables
    sum_distances = 0        # Accumulator for the sum of distances
    pair_count = 0           # Counter for the number of unique pairs

    # Calculate pairwise Hamming distances
    FOR each individual i in Population P (i = 1 to N)
        FOR each individual j in Population P (j = i+1 to N)

            # Compute Hamming distance
            hamming_distance = 0    # Initialize distance for the current pair
            FOR bit_position k = 1 to L
                IF P[i][k] not equal to P[j][k] THEN
                    hamming_distance = hamming_distance + 1
                END IF
            END FOR

            # Accumulate the distance
            sum_distances = sum_distances + hamming_distance

            # Increment the pair count
            pair_count = pair_count + 1
```

```
        END FOR
    END FOR

    # Compute average diversity
    average_diversity = sum_distances / pair_count

    # Output the result
    RETURN average_diversity

END Calculate Diversity based on the Hamming Distance
```

Pseudocode explanation

- Input Parameters: The function defines the number of solutions (N) and the number of decision variables (D). The binary length of each solution is computed in (L). The variables sum_distances and pair_count are initialized to sum the distances of the solutions and count the total unique pairs of solutions, respectively.
- Calculating the diversity: A nested pair of loops runs n times over each solution pair. For each iteration, the Hamming distance is calculated according to the differences between bit positions. In turn, the variables sum_distances and pair_count are updated to the actual value.
- Result: Once all n solutions pairs have been analyzed through the Hamming distance diversity method, the average_diversity value is assessed by dividing the sum_distances with the pair_count, and finally, the average diversity result based on Hamming distance calculation is returned.

7.2.2.2 MATLAB Code

```
function avgDiversity = HammingDistance(population)
    % HammingDistance = Calculates the average diversity of a population
    % based on Hamming distance.
    % population - Matrix (NxL) where N is the number of solutions and
    %              L is the length of each binary solution.

    % Population size
    numSolutions = size(population, 1); % Number of solutions
    solutionLength = size(population, 2); % Length of each solution

    % Check that all solutions have the same length
    if any(size(population, 2) ~= solutionLength)
        error('All solutions must have the same length!');
    end

    % Initialize the sum of distances and pair counter
    totalDistance = 0;
    pairCount = 0;

    % Calculate the Hamming distances between each pair of solutions
    for i = 1:numSolutions
        for j = i+1:numSolutions
            % Calculate Hamming distance between solution i and j
            hammingDistance = sum(population(i, :) ~= population(j, :));
            totalDistance = totalDistance + hammingDistance; % Sum distances
            pairCount = pairCount + 1;                       % Pair Counting
        end
    end
```

```
    % Calculate average diversity
    avgDiversity = totalDistance / pairCount;

    % Display results
    fprintf('The average diversity based on the Hamming distance is: %.2f\n',
     avgDiversity);
end
```

Example input for testing the code in MATLAB

```
% Define a population vector of 6 individuals with binary representation length
population = [
    1 0 1 1 0 1;
    0 1 0 1 1 0;
    1 1 1 0 0 1;
    0 0 1 1 1 0;
    1 0 0 0 1 1;
    0 1 1 0 0 0
];

% Call the function
avgDiversity = HammingDistance(population);
```

7.2.3 *Average Population Diversity*

Introduced by Poláková et al. [7], APD assesses population diversity based on individual characteristics. It serves as an adaptive population size mechanism in Differential Evolution (DE). Equation 7.5 defines d_{AP}, estimating diversity via mean squared distance between individuals and the population mean per dimension.

$$d_{AP} = \sqrt{\frac{1}{N}\sum_{i=1}^{N}\sum_{j=1}^{D}(x_{ij} - \bar{x_j})^2} \;\;, where, \; \bar{x_j} = \frac{1}{N}\sum_{i=1}^{N} x_{ij} \tag{7.5}$$

Equation 7.5 defines N as the number of candidate solutions and D as the number of dimensions. x_{ij} represents the j_{th} dimension value for the i_{th} individual, while $\bar{x_j}$ is the population mean in dimension j. The centroid $\bar{x} = (\bar{x}_1, \bar{x}_2, ..., \bar{x}_D)$ estimates population spread relative to its average position. To illustrate, consider a population of $N = 5$ individuals in $D = 2$ dimensions. Their coordinates are described in matrix X, where each row represents an individual, and each column depicts a dimension.

$$X = \begin{bmatrix} 1.00 & 2.50 \\ 1.50 & 1.80 \\ 0.90 & 2.20 \\ 1.20 & 3.10 \\ 1.30 & 1.90 \end{bmatrix}$$

The centroid analysis is first estimated using the equation for $\bar{x_j}$ with $j = 1, 2$ based on Equation 7.5. The calculation is as follows:

$$\bar{x_1} = \frac{1.00 + 1.50 + 0.90 + 1.20 + 1.30}{5} = \frac{5.9}{5} = 1.18$$

$$\bar{x}_2 = \frac{2.50 + 1.80 + 2.20 + 3.10 + 1.90}{5} = \frac{11.50}{5} = 2.30$$

$$\bar{x} = (1.18, 2.30)$$

Once the centroid $(1.18, 2.30)$ is obtained, the following phase is to assess the squared distances $(x_{ij} - \bar{x}_j)^2$ and sum these distances for each individual i within the population.

- **Individual 1:** $x_1 = [1.00 \;\; 2.50]$

$$(x_{11} - \bar{x}_1)^2 = (1.00 - 1.18)^2 = (-0.18)^2 = 0.0324$$

$$(x_{12} - \bar{x}_2)^2 = (2.50 - 2.30)^2 = (0.20)^2 = 0.0400$$

$$x_1 = 0.0324 + 0.0400 = 0.0724$$

- **Individual 2:** $x_2 = [1.50 \;\; 1.80]$

$$(x_{21} - \bar{x}_1)^2 = (1.50 - 1.18)^2 = (0.32)^2 = 0.1024$$

$$(x_{22} - \bar{x}_2)^2 = (1.80 - 2.30)^2 = (-0.50)^2 = 0.2500$$

$$x_2 = 0.1024 + 0.2500 = 0.3524$$

- **Individual 3:** $x_3 = [0.90 \;\; 2.20]$

$$(x_{31} - \bar{x}_1)^2 = (0.90 - 1.18)^2 = (-0.28)^2 = 0.0784$$

$$(x_{32} - \bar{x}_2)^2 = (2.20 - 2.30)^2 = (-0.10)^2 = 0.0100$$

$$x_3 = 0.0784 + 0.0100 = 0.0884$$

- **Individual 4:** $x_4 = [1.20 \;\; 3.10]$

$$(x_{41} - \bar{x}_1)^2 = (1.20 - 1.18)^2 = (0.02)^2 = 0.0004$$

$$(x_{42} - \bar{x}_2)^2 = (3.10 - 2.30)^2 = (0.80)^2 = 0.6400$$

$$x_4 = 0.0004 + 0.6400 = 0.6404$$

- **Individual 5:** $x_5 = [1.30 \;\; 1.90]$

$$(x_{51} - \bar{x}_1)^2 = (1.30 - 1.18)^2 = (0.12)^2 = 0.0144$$

$$(x_{52} - \bar{x}_2)^2 = (1.90 - 2.30)^2 = (-0.40)^2 = 0.1600$$

$$x_5 = 0.0144 + 0.1600 = 0.1744$$

The next step is to sum all the previously calculated squared distances for each candidate solution ($x_1, ..., x_5$):

$$\sum_{i=1}^{5} = 0.0724 + 0.3524 + 0.0884 + 0.6404 + 0.1744 = 1.3280$$

Finally, with this information, the calculation of the average population diversity d_{AP} is as under:

$$d_{AP} = \sqrt{\frac{1}{5}1.3280} = \sqrt{0.2656} \approx 0.5154$$

Figure 7.4 visually depicts individual distribution and distances to the centroid. Black circles represent population members, while the hexagon marks the centroid. Dashed lines illustrate distances. The average population diversity (d_{AP}) is 0.5154, indicating moderate diversity—higher values imply greater variation, while lower (≈ 0.1) suggests concentration. A high d_{AP} supports exploration in early optimization, whereas low d_{AP} signals convergence. A rapid decline may indicate premature convergence, trapping the algorithm in a suboptimal region.

The APD metric quantifies population dispersion by measuring distances between individuals and the centroid in feature space. Squared differences with the population mean assess search space exploration and convergence rate—a rapid decline may hinder finding the global optimum. Pseudocode and MATLAB code are detailed in Subsections 7.2.3.1 and 7.2.3.2.

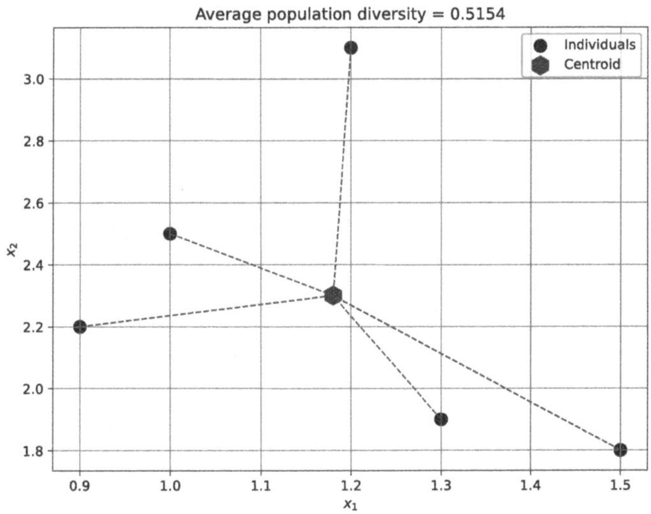

Figure 7.4: Representation of the individuals and their distances to the centroid.

7.2.3.1 Pseudocode

```
BEGIN Calculate Average Population Diversity

    # Define the problem
    Define the population size N
    Define the dimensionality of the search space S

    # Initialize variables
    centroid = [0] * S    # Vector to store the mean of each dimension
    sum_squared_distances = 0    # Accumulator for squared distances

    # Compute the centroid of the population
    FOR each dimension j in search space (j = 1 to S)
        FOR each individual i in Population P (i = 1 to N)
            centroid[j] = centroid[j] + P[i][j]    # Sum values in each dimension
        END FOR
        centroid[j] = centroid[j] / N    # Compute mean for dimension j
    END FOR

    # Compute the squared distances from the centroid
    FOR each individual i in Population P (i = 1 to N)
        FOR each dimension j in search space (j = 1 to S)
            squared_distance = (P[i][j] - centroid[j])^2    # Compute squared
     difference
            sum_squared_distances = sum_squared_distances + squared_distance    #
      Accumulate
        END FOR
    END FOR

    # Compute the final Average Population Diversity
    APD = SQRT(sum_squared_distances / N)

    # Output the result
    RETURN APD

END Calculate Average Population Diversity
```

Pseudocode explanation

- Input Parameters: The function defines the number of solutions (N) and the dimensions of the search space (S). The binary length of each solution is computed in (L). The variable sum_squared_distances is initialized to sum the squared distances of the solutions. A centroid is initialized to keep the mean of each dimension of the problem.
- Calculating the diversity: A nested pair of loops runs n times over each dimension and solution in order to assess the centroid of the population. Once the centroid value is obtained, a nested pair of loops run n times over each dimension and solution to assess the sum_squared_distances based on the centroid.
- Result: Once all n solutions have been analyzed, the Average population diversity is assessed by calculating the square root of sum_squared_distances divided by N. Finally, the average population diversity result is returned.

7.2.3.2 MATLAB Code

```
function avgDiversity = AveragePopulationDiversity(population)
    % AveragePopulationDiversity - Calculates the average population
    % diversity (APD) based on the squared distance from the
    % population mean.
    % population - Matrix (NxD) where N is the number of solutions
    %              and D is the number of dimensions (features).

    % Population size and number of dimensions
    [numSolutions, numDimensions] = size(population);

    % Compute the centroid (mean value per dimension)
    centroid = mean(population, 1);

    % Initialize the sum of squared distances
    totalSquaredDistance = 0;

    % Compute squared distances for each individual
    for i = 1:numSolutions
        squaredDistance = sum((population(i, 1:numDimensions) - centroid(1:
     numDimensions)).^2);
        totalSquaredDistance = totalSquaredDistance + squaredDistance;
    end

    % Compute the average population diversity (APD)
    avgDiversity = sqrt(totalSquaredDistance / numSolutions);

    % Display the result
    fprintf('The average population diversity (APD) is: %.4f\n', avgDiversity);
end
```

Example input for testing the code in MATLAB

```
% Define a population matrix of 10 individuals in a 5-dimensional space
population = [
    2, 4, 6, 8, 10;
    5, 1, 3, 7, 9;
    7, 8, 2, 4, 6;
    4, 9, 5, 3, 1;
    1, 6, 7, 2, 8;
    8, 3, 9, 5, 0;
    0, 2, 4, 1, 7;
    6, 7, 1, 9, 3;
    3, 5, 8, 0, 2;
    9, 0, 2, 6, 4
];

% Call the function
avgDiversity = AveragePopulationDiversity(population);
```

7.3 Entropy-based Methods

First introduced in physics, entropy quantifies the number of microstates corresponding to a macrostate [12]. Systems with greater movement possibilities exhibit high entropy, whereas rigidly fixed particles have low entropy. For instance, frozen water has low entropy due to its structured molecules. As ice melts, molecular movement increases, raising entropy. In a gaseous state, molecules move most freely, resulting in high entropy.

Diversity mechanisms fall into distance-based, entropy-based, and variance-based categories. This section highlights an entropy-based approach, where entropy quantifies uncertainty and disorder across fields like information theory, physics, and chemistry [13]. Given the computational focus of this book, we adopt the formal definition from information theory. While various entropy measures exist [14], Shannon entropy is explored for general understanding.

7.3.1 *Shannon Entropy*

In 1948, Claude Shannon adapted entropy for information theory, linking knowledge to state probability [15]. Greater predictability reduces information content (low entropy), while unpredictability increases it (high entropy). Figure 7.5 illustrates this concept using three boxes with heads/tails coin arrangements.

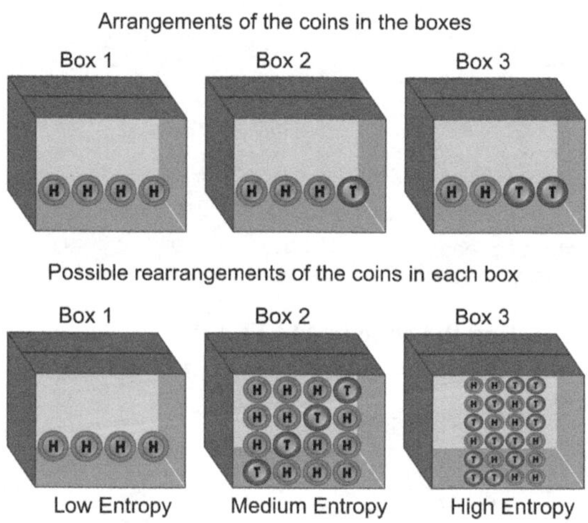

Figure 7.5: Arrangements and entropy of boxes with heads and tails coins.

Entropy Levels in Fig. 7.5:

Box 1 has low entropy due to high predictability (100% heads). Box 2 offers medium entropy with mixed probabilities (75% heads, 25% tails). Box 3, being highly unpredictable (50% heads, 50% tails), exhibits high entropy. Greater species variety increases entropy further.

The formal definition for the Shannon entropy is provided in Equation 7.6. This formulation enables the measurement of uncertainty or entropy in the species distribution (solutions) within a given population. From Equation 7.6, p_k represents the relative abundance of solutions k in the population, and N is the total number of individuals in the population.

$$d_{SE} = -\sum_{k=1}^{N} p_k \log(p_k) \tag{7.6}$$

As can be observed, the term $p_k \log(p_k)$ depicts the weighted contribution of each unique solution k to the overall diversity. To calculate the relative abundance p_k, each unique solution frequency f_k is divided for N as $p_k = \frac{f_k}{N}$. The negative sign ensures the index gives a positive value, as the logarithmic term is negative for probabilities less than 1. The logarithm for Shannon Entropy is normally established as base 2 or base e. Notably, the greater the value of d_{SE}, the higher the diversity, which signifies a more uniform distribution of individuals across the solutions. Contrarily, a lower value of d_{SE} indicates that few solutions dominate the population, which suggests lower diversity. The following example is intended to improve the understanding of the Shannon entropy concept and its application in evolutionary algorithms. To develop this example, we will take the following population of 25 individuals in a 2-dimensional space:

$$\text{population} = \left\{ \begin{array}{lllll} (1.0, 2.0), & (1.0, 2.0), & (1.0, 2.0), & (1.0, 2.0), & (1.0, 2.0), \\ (2.5, 3.8), & (2.5, 3.8), & (2.5, 3.8), & (2.5, 3.8), & (3.1, 4.2), \\ (3.1, 4.2), & (3.1, 4.2), & (3.1, 4.2), & (0.5, 1.5), & (0.5, 1.5), \\ (0.5, 1.5), & (4.0, 2.5), & (4.0, 2.5), & (3.5, 2.0), & (3.5, 2.0), \\ (1.8, 4.0), & (2.2, 1.8), & (3.0, 3.5), & (1.7, 2.3), & (2.0, 3.0) \end{array} \right\}$$

From the individuals, it can be observed that multiple repeated solutions exist, which is desirable since, in evolutionary algorithms, the solutions tend to be similar when converging to a local optimum or global optimum. So, the first step is to analyze the existing number of unique solutions and their relative abundance p_k. This information is summarized in Table 7.3.

Table 7.3: Relative abundance p_k of each unique solution k.

Solution	Frequency	Relative Abundance
(1.0, 2.0)	5	$5/25 = 0.20$
(2.5, 3.8)	4	$4/25 = 0.16$
(3.1, 4.2)	4	$4/25 = 0.16$
(0.5, 1.5)	3	$3/25 = 0.12$
(4.0, 2.5)	2	$2/25 = 0.08$
(3.5, 2.0)	2	$2/25 = 0.08$
(1.8, 4.0)	1	$1/25 = 0.04$
(2.2, 1.8)	1	$1/25 = 0.04$
(3.0, 3.5)	1	$1/25 = 0.04$
(1.7, 2.3)	1	$1/25 = 0.04$
(2.0, 3.0)	1	$1/25 = 0.04$

The relative abundance of each unique solution is taken from Table 7.3 and applied in the Shannon entropy Equation 7.6 to perform the corresponding summation, as follows:

$$d_{SE} = -[(0.20 \times \log_2(0.20)) + 2 \times (0.16 \times \log_2(0.16)) + (0.12 \times \log_2(0.12)) + 2 \times (0.08 \times \log_2(0.08)) + 5 \times (0.04 \times \log_2(0.04))]$$

$$d_{SE} = -[(0.20 \times -2.3219) + 2 \times (0.16 \times -2.6439) + (0.12 \times -3.0589) + 2 \times (0.08 \times -3.6439) + 5 \times (0.04 \times -4.6439)]$$

$$d_{SE} = -[-0.4644 - 0.8460 - 0.3671 - 0.5830 - 0.9288]$$

$$d_{SE} = -[-3.1893] = 3.1893$$

The graphical representation of the population solution points and their relative abundances, as well as the Shannon entropy result, are presented in Fig. 7.6. The result obtained for the diversity criterion based on Shannon entropy is $d_{SE} = 3.1893$. This result implies that the population is diverse and has a moderately high entropy but is starting to converge. The maximum possible entropy would be obtained if all the species were different from each other, i.e., if all the 25 solutions of the population were located at different points, the entropy would be $log_2(25) = 4.6439$. If the result obtained were that, then the maximum diversity could be assured. On the other hand, if the entropy result falls below 1, this would indicate that 1 or 2 solutions dominate the solutions and, therefore, require a mechanism to improve the diversity.

The Shannon entropy has been identified as a particularly valuable metric for assessing the evenness and richness of a population [16]. Tracking the Shannon entropy index over time makes it possible to monitor how the population's diversity evolves, thereby ensuring a healthy balance between exploration and exploitation. In turn, Shannon entropy can be extended to more complex scenarios, such as in continuous spaces, by treating different solution regions as "species" and applying the index to measure the spread of solutions across the entire search space. In such contexts, the Shannon entropy index emerges as a reliable mechanism for evaluating the diversity of a population in terms of the distribution of solutions across the solution space, thereby enhancing the efficacy of decision-making in optimization processes. The pseudocode and the MATLAB code for the Shannon Entropy diversity metric are presented in Subsections 7.3.1.1 and 7.3.1.2, respectively.

7.3.1.1 Pseudocode

```
BEGIN Calculate Shannon Entropy

    # Define the problem
    Define the population size N
    Define the dimensionality of the search space S
```

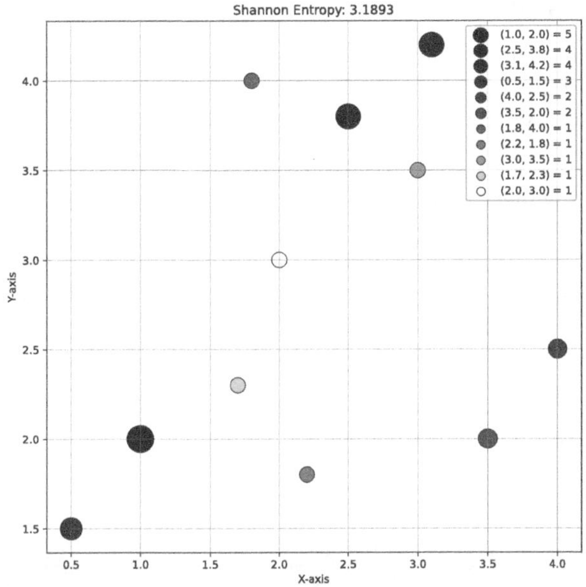

Figure 7.6: Shannon entropy and relative abundance of each unique solution.

```
    # Initialize variables
    unique_solutions = [] # Vector to store unique solutions
    occurrences = [] # Vector to store occurrences of each unique solution
    entropy = 0 # Shannon entropy accumulator

    # Count occurrences of each unique solution in the population
    FOR each individual i in Population P (i = 1 to N)
        found = FALSE
        FOR each index j in (1 to LENGTH(unique_solutions))
            IF P[i] == unique_solutions[j] THEN
                occurrences[j] = occurrences[j] + 1
                found = TRUE
                BREAK
            END IF
        END FOR
        IF found == FALSE THEN
            ADD P[i] to unique_solutions
            ADD 1 to occurrences
        END IF
    END FOR

    # Compute Shannon Entropy
    FOR each index k in (1 to LENGTH(unique_solutions))
        probability = occurrences[k] / N # Compute relative abundance
        entropy = entropy - (probability * LOG(probability , 2)) # Shannon
     entropy formula
    END FOR

    # Output the result
    RETURN entropy

END Calculate Shannon Entropy
```

Pseudocode explanation

- Input Parameters: The function sets the number of solutions (N), defining vectors $\text{unique}_{\text{solutions}}$ and occurrences to store each unique solution and its frequency. The entropy variable is initialized to accumulate the entropy calculation.
- Calculating the diversity: A loop scans all solutions to identify unique entries and count occurrences. A nested loop verifies whether an individual is already recorded—if found, its count updates; otherwise, it is added as new. Once frequencies are determined, Shannon's entropy formula computes entropy, where probabilities are derived from occurrence counts divided by N, and the weighted sum of $p_k \log_2(p_k)$ is accumulated for the final entropy value.
- Result: After processing all N solutions, the Shannon entropy value is returned, representing the diversity of the population.

7.3.1.2 MATLAB Code

```
function entropyValue = ShannonEntropy(population)
    % ShannonEntropy - Calculates the Shannon entropy of a population
    % as a measure of diversity.
    % population - Matrix (NxD) where N is the number of particles and
    %                D is the dimensionality of space.

    % Population Size
    numParticles = size(population, 1);

    % Initialize occurrences
    occurrences = zeros(numParticles, 1); % Preallocate for the occurrences

    % Number of unique solutions found
    numUnique = 0;

    % Identify unique solutions and count occurrences
    for i = 1:numParticles
        found = false;
        for j = 1:numUnique
            if isequal(population(i, :), population(j, :)) % Compare directly in
  population
                occurrences(j) = occurrences(j) + 1;
                found = true;
                break;
            end
        end
        if ~found
            numUnique = numUnique + 1;
            population(numUnique, :) = population(i, :); % Store the unique
 solution in place
            occurrences(numUnique) = 1; % Initialize occurrence for the new
 unique solution
        end
    end

    % Truncate the unused portion of the occurrences
    occurrences = occurrences(1:numUnique);

    % Compute Shannon entropy
    entropyValue = 0;
    for k = 1:numUnique
        pk = occurrences(k) / numParticles;
```

```
        entropyValue = entropyValue - pk * log2(pk);
    end

    % Display results
    fprintf('The Shannon entropy of the population is: %.4f\n', entropyValue);
end
```

Example input for testing the code in MATLAB

```
% Define a population matrix of 10 individuals in a 3-dimensional space
population = [
    1, 2, 3;
    4, 5, 6;
    1, 2, 3;
    7, 8, 9;
    4, 5, 6;
    1, 2, 3;
    10, 11, 12;
    9, 7, 5;
    10, 11, 12;
    1, 2, 3;
];

% Call the function
entropyValue = ShannonEntropy(population);
```

7.4 Variance-based Methods

Variance quantifies data dispersion around the arithmetic mean, aiding insights in economics, engineering, and sciences. High variance indicates greater spread, while low variance suggests data concentration around the mean. Figure 7.7 illustrates this, showing two sets of 8,000 samples—both with a mean of 100, but Set 1 (variance = 1,000) is tightly clustered, whereas Set 2 (variance = 10,000) is widely spread.

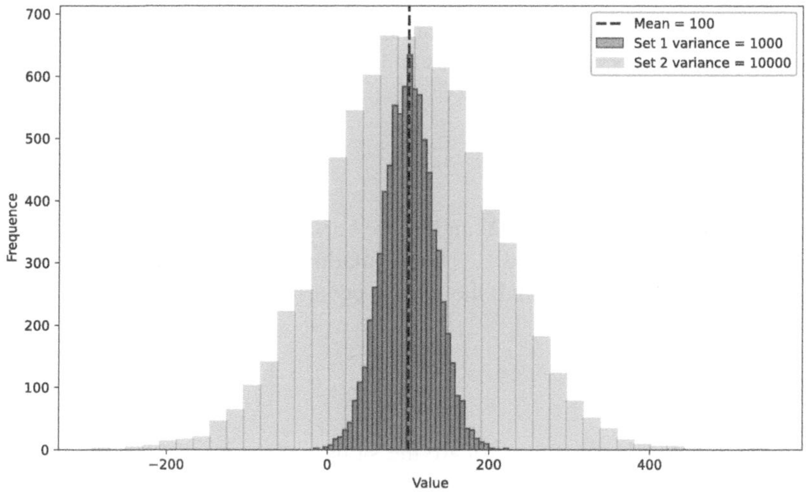

Figure 7.7: Difference in the variance of two sets with the same mean.

Variance is evaluated for either a full population or a sample, depending on data availability. Population variance applies when all data is accessible, while sampling variance estimates population dispersion using a subset. Equation 7.7 presents the formula for computing population (P) variance.

$$Variance_P = \frac{1}{N}\sum_{i=1}^{N}(x_i - \mu)^2 \tag{7.7}$$

where N is the total number of elements in a population, x_i is each individual value in the population, and μ is the population mean ($\mu = \frac{1}{N}\sum_{i=1}^{N} x_i$). The sampling ($S$) variance is assessed by:

$$Variance_S = \frac{1}{n-1}\sum_{i=1}^{n}(x_i - \overline{x})^2 \tag{7.8}$$

where n is the sample size, x_i is each individual value in the sample, and $\overline{x}$ is the sample mean ($\overline{x} = \frac{1}{n}\sum_{i=1}^{n} x_i$). The term $n-1$ is employed to fix the bias in estimating population variance from a sample (known as Bessel correction). Both population and sampling variances provide insights into the population's variability concerning each solution feature and how solutions differ regarding their collective characteristics. Thus, variance measurement is a valuable tool for assessing diversity in evolutionary algorithms [17]. This section will focus on a specific measure of variance that has been employed to assess diversity. This measure is referred to as the covariance matrix.

7.4.1 *Covariance Matrix*

Covariance quantifies dependence between two random variables, forming the basis for linear correlation and regression. It is computed as population or sample covariance, depending on data availability. The population (P) covariance measures the linear relationship between two variables in a whole population and is estimated as:

$$Covariance_P = \frac{1}{N}\sum_{i=1}^{N}(x_i - \mu_X)(y_i - \mu_Y) \tag{7.9}$$

where N is the population size, X and Y are variables. x_i and y_i are the values of observation i on variables X and Y. The population means of X and Y are depicted as μ_X and μ_Y. On the other hand, the sample covariance is formulated as follows:

$$Covariance_S = \frac{1}{n-1}\sum_{i=1}^{n}(x_i - \overline{x})(y_i - \overline{y}) \tag{7.10}$$

where n is the sample size, $\overline{x}$ and $\overline{y}$ are the sample means, and $n-1$ is the Bessel correction. If X is a random n-dimensional vector, then the covariance is transformed into a matrix.

The covariance matrix, also called the variance-covariance matrix, is utilized when it is crucial to understand the correlation between multiple variables. In the matrix, the main diagonal represents the variance of each variable, and the covariances between variable pairs are denoted outside the diagonal. The generalization of an $n \times n$ dimensional covariance matrix $\sum$ is depicted as follows:

$$\sum = \begin{bmatrix} \mathbf{Var(X_1)} & Cov(X_1, X_2) & \cdots & Cov(X_1, X_n) \\ Cov(X_2, X_1) & \mathbf{Var(X_2)} & \cdots & Cov(X_2, X_n) \\ \vdots & \vdots & \ddots & \vdots \\ Cov(X_n, X_1) & Cov(X_n, X_2) & \cdots & \mathbf{Var(X_n)} \end{bmatrix} \tag{7.11}$$

In Equation 7.11, the variance is represented by $Var(X_n)$, and the main diagonal is highlighted in **bold**. The covariance is illustrated as a pair of solutions $Cov(X_i, X_n)$; as mentioned above, its solutions lie outside the main diagonal. The covariance matrix presents the following properties:

- The number of columns and rows of the matrix is equal. So, the covariance matrix is squared.
- The matrix is symmetric, so the transpose matrix equals the original.
- The covariance matrix is positive semi-definite if the variables are not linearly dependent.

Let us look at an example to understand the covariance matrix concept better. Suppose we run an evolutionary algorithm to optimize an objective function $f(x)$ in a 3-dimensional space (x_1, x_2, x_3). After several generations, we have the following individuals in the population:

Table 7.4: Collected data.

Individual	x_1	x_2	x_3
1	2.0	3.0	1.5
2	2.5	3.2	2.0
3	3.0	3.5	2.5
4	4.0	4.2	3.8
5	4.5	4.8	4.2

In the data context, the population size is $N = 5$. First we calculate the mean of each variable $\bar{x_1}, \bar{x_2}$, and $\bar{x_3}$ as:

$$\bar{x_1} = \frac{2.0 + 2.5 + 3.0 + 4.0 + 4.5}{5} = 3.20$$
$$\bar{x_2} = \frac{3.0 + 3.2 + 3.5 + 4.2 + 4.8}{5} = 3.74$$
$$\bar{x_3} = \frac{1.5 + 2.0 + 2.5 + 3.8 + 4.2}{5} = 2.80$$

With the mean, the next step is to estimate the respective variance of x_1, x_2, and x_3:

$$var_{x_1} = \frac{(2 - 3.2)^2 + (2.5 - 3.2)^2 + (3 - 3.2)^2 + (4 - 3.2)^2 + (4.5 - 3.2)^2}{4}$$
$$var_{x_1} = \frac{(-1.2)^2 + (-0.7)^2 + (-0.2)^2 + (0.8)^2 + (1.3)^2}{4}$$
$$var_{x_1} = \frac{1.44 + 0.49 + 0.04 + 0.64 + 1.69}{4} = \frac{4.3}{4} = 1.075$$

$$var_{x_2} = \frac{(3 - 3.7)^2 + (3.2 - 3.7)^2 + (3.5 - 3.7)^2 + (4.2 - 3.7)^2 + (4.8 - 3.7)^2}{4}$$
$$var_{x_2} = \frac{(-0.74)^2 + (-0.54)^2 + (-0.24)^2 + (0.46)^2 + (1.06)^2}{4}$$
$$var_{x_2} = \frac{0.5476 + 0.2916 + 0.0576 + 0.2116 + 1.1236}{4} = \frac{2.232}{4} = 0.558$$

$$var_{x_3} = \frac{(1.5 - 2.8)^2 + (2 - 2.8)^2 + (2.5 - 2.8)^2 + (3.8 - 2.8)^2 + (4.2 - 2.8)^2}{4}$$
$$var_{x_3} = \frac{(-1.3)^2 + (-0.8)^2 + (-0.3)^2 + (1.0)^2 + (1.4)^2}{4}$$
$$var_{x_3} = \frac{1.69 + 0.64 + 0.09 + 1 + 1.96}{4} = \frac{5.38}{4} = 1.345$$

The next step assesses covariance for all possible pairs (x_1, x_2), (x_1, x_3), and (x_2, x_3) using Equation 7.9. To minimize redundancy, variance values computed for each individual are utilized:

$$cov_{x_1,x_2} = \frac{1}{4}[(-1.2)(-0.74) + (-0.7)(-0.54) + (-0.2)(-0.24) + (0.8)(0.46) + (1.3)(1.06)]$$

$$cov_{x_1,x_2} = \frac{0.888 + 0.378 + 0.048 + 0.368 + 1.378}{4} = \frac{3.06}{4} = 0.765$$

$$cov_{x_1,x_3} = \frac{1}{4}[(-1.2)(-1.3) + (-0.7)(-0.8) + (-0.2)(-0.3) + (0.8)(1) + (1.3)(1.4)]$$

$$cov_{x_1,x_3} = \frac{1.56 + 0.56 + 0.06 + 0.8 + 1.82}{4} = \frac{4.8}{4} = 1.2$$

$$cov_{x_2,x_3} = \frac{1}{4}[(-0.74)(-1.3) + (-0.54)(-0.8) + (-0.24)(-0.3) + (0.46)(1) + (1.06)(1.4)]$$

$$cov_{x_2,x_3} = \frac{0.962 + 0.432 + 0.072 + 0.46 + 1.484}{4} = \frac{3.41}{4} = 0.8525$$

In the last step, the covariance matrix is formed. Since there are 3 variables, the matrix would be 3×3:

$$\sum = \begin{bmatrix} \mathbf{1.0750} & 0.7650 & 1.2000 \\ 0.7650 & \mathbf{0.5580} & 0.8525 \\ 1.2000 & 0.8525 & \mathbf{1.3450} \end{bmatrix}$$

From the covariance matrix, it can be interpreted that of the variances described in the main diagonal, the dimension x_3 is the one with the highest dispersion in its values; the values found in x_2 are not very dispersed and are concentrated around the mean, and x_1 has a moderate dispersion. Furthermore, the covariances described around the main diagonal show that when x_1 increases, x_2 will also increase. While x_1 and x_3 have a strong positive correlation and, therefore, will vary together. Finally, x_2 and x_3 have a significant relationship but not as strong as that of x_1 with x_3.

This information is valuable when implementing the covariance matrix as a diversity strategy in evolutionary algorithms since it can provide an analysis of the structural diversity of the population and detect correlations among decision variables [18]. The covariance matrix reveals population distribution in the search space. Variance indicates whether the algorithm is in exploration (high variance) or convergence (low variance). High covariance suggests dimension correlation, influencing crossover and mutation selections. Pseudocode and MATLAB code are in Subsections 7.4.1.1 and 7.4.1.2.

7.4.1.1 Pseudocode

```
BEGIN Calculate Covariance Matrix

# Define the problem
Define the population size N
Define the dimensionality of the search space S

# Initialize variables
Mean_Vector = [0] * S  # Vector to store mean values of each dimension
Cov_Matrix = [[0] * S for each dimension]  # Initialize covariance matrix

# Compute mean values for each dimension
FOR each dimension d in (1 to S)
    Sum_d = 0
    FOR each individual i in Population P (i = 1 to N)
        Sum_d = Sum_d + P[i][d]
    END FOR
    Mean_Vector[d] = Sum_d / N
END FOR

# Compute covariance matrix
FOR each dimension i in (1 to S)
    FOR each dimension j in (1 to S)
        Sum_Cov = 0
        FOR each individual k in Population P (k = 1 to N)
            Sum_Cov = Sum_Cov + (P[k][i] - Mean_Vector[i]) * (P[k][j] -
    Mean_Vector[j])
        END FOR
        Cov_Matrix[i][j] = Sum_Cov / (N - 1)  # Compute covariance
    END FOR
END FOR

# Output the result
RETURN Cov_Matrix

END Calculate Covariance Matrix
```

Pseudocode explanation

- Input Parameters: The function defines the population size (N) and the dimension of the search space (S). The vector Mean_Vector of size S is initialized to store the information of the mean values for each dimension. The Cov_Matrix is a matrix initialized with zeros.
- Calculating the diversity: The first loop iterates over all dimensions to compute the mean value of each variable across the population. Next, a nested loop iterates over all pairs of dimensions (i, j) to compute the covariance matrix. For each pair, a summation is performed over all individuals, accumulating the product of the deviations of the values from their respective mean values. The final covariance for each pair (i, j) is obtained by dividing by $N - 1$, ensuring an unbiased estimate.

- Result: After processing all N solutions, the Covariance matrix is returned, representing the diversity of the population.

7.4.1.2 MATLAB Code

```
function covMatrix = CovarianceMatrix(population)
    % CovarianceMatrix - Calculates the covariance matrix of a population
    % as a measure of diversity in evolutionary algorithms.
    % population - Matrix (NxD) where N is the number of individuals and
    %              D is the dimensionality of the search space.

    % Population Size and Dimensionality
    [numParticles, numDimensions] = size(population);

    % Compute Mean Vector for Each Dimension
    meanVector = mean(population, 1);

    % Initialize Covariance Matrix
    covMatrix = zeros(numDimensions, numDimensions);

    % Compute Covariance Matrix
    for i = 1:numDimensions
        for j = 1:numDimensions
            sumCov = 0;
            for k = 1:numParticles
                sumCov = sumCov + (population(k, i) - meanVector(i)) * (
   population(k, j) - meanVector(j));
            end
            covMatrix(i, j) = sumCov / (numParticles - 1); %numParticles-1 =
   Bessel correction
        end
    end

    % Display Results
    fprintf('Covariance Matrix of the Population:\n');
    disp(covMatrix);
end
```

Example input for testing the code in MATLAB

```
% Define a population matrix of 10 individuals in a 5-dimensional space
population = [
    2.1, 3.2, 1.5, 4.0, 2.8;
    2.5, 3.1, 2.0, 4.1, 3.0;
    3.0, 3.5, 2.8, 4.3, 3.5;
    4.2, 4.0, 3.5, 4.7, 3.9;
    4.8, 4.5, 4.0, 5.0, 4.2;
    2.2, 3.0, 1.8, 4.0, 2.9;
    3.5, 3.8, 2.9, 4.5, 3.8;
    4.0, 4.2, 3.7, 4.8, 4.0;
    5.0, 4.9, 4.5, 5.5, 4.5;
    3.8, 3.7, 3.0, 4.4, 3.6
];

% Call the function
covMatrix = CovarianceMatrix(population);
```

7.5 Alternative Metrics for Measuring Diversity

Beyond distance-based, entropy-based, and variance-based diversity measures, alternative metrics integrate strategies for a broader diversity assessment. These methods analyze population structure and provide insights at various evolutionary stages.

Several approaches have been proposed in literature [19, 20]; this section highlights Dimension-Wise Diversity, exploring its mathematical foundations and potential applications.

7.5.1 *Dimension-wise Diversity*

Evolutionary algorithms use candidate solutions to explore the search space. The best-performing search agents guide optimization, reducing distance and increasing exploitation. Greater separation enhances exploration. To quantify these changes, the Dimension-Wise Diversity (DWD) metric [21] is applied. This metric defines population diversity as follows:

$$d_j = \frac{1}{n}\sum_{i=1}^{n}|\text{median}(x_j) - x_{ij}| \quad (7.12)$$

$$d_{DWD} = \frac{1}{S}\sum_{j=1}^{S} d_j \quad (7.13)$$

In Equations 7.12 and 7.13, $median(x_j)$ represents the median of dimension j across the entire population, x_{ij} is the value of dimension j for search agent i, N is the number of search agents in the population, and S is the number of design variables (dimensions) in the optimization problem.

The diversity in each dimension d_j is defined as the average distance between the dimension j of every search agent and the median of that dimension. The overall dimension-wise population diversity d_{DWD} is calculated by averaging d_j across all dimensions. In evolutionary algorithms, these values are computed at every iteration. To evaluate the balance between exploration and exploitation, the following models are used:

$$ExplorationRate = \left(\frac{d_{DWD}}{d_{max}}\right) \times 100 \quad (7.14)$$

$$ExploitationRate = \left(\frac{|d_{DWD} - d_{max}|}{d_{max}}\right) \times 100 \quad (7.15)$$

where d_{max} represents the maximum diversity value found during optimization. The percentage of exploration $ExplorationRate$ indicates the level of exploration relative to the maximum diversity, while the percentage of exploitation $ExploitationRate$ represents the level of exploitation as the complementary percentage. Both metrics are mutually conflicting and complementary. For a better understanding of these concepts of exploration and exploitation rates, let us set two population examples of each state, where the number of search agents is $N = 10$, and the number of design variables is $S = 3$:

Table 7.5 illustrates a population in exploration phase, where agents vary widely, resulting in high diversity measures. Table 7.6, in contrast, depicts the same search

Table 7.5: Search agents in a exploration phase.

Agent	x_1	x_2	x_3
1	4.3709	9.5564	7.5879
2	6.3879	2.4042	2.4040
3	1.5228	8.7956	6.4100
4	7.3727	1.1853	9.7292
5	8.4920	2.9111	2.6364
6	2.6506	3.7382	5.7228
7	4.8875	3.6211	6.5067
8	2.2554	3.6293	4.2973
9	5.1046	8.0666	2.7971
10	5.6281	6.3317	1.4181

Table 7.6: Search agents in a exploitation phase.

Agent	x_1	x_2	x_3
1	3.6075	3.1705	3.0651
2	3.9489	3.9656	3.8084
3	3.3046	3.0977	3.6842
4	3.4402	3.1220	3.4952
5	3.0344	3.9093	3.2588
6	3.6625	3.3117	3.5201
7	3.5467	3.1849	3.9696
8	3.7751	3.9395	3.8948
9	3.5979	3.9219	3.0885
10	3.1960	3.0452	3.3253

agents in exploitation phase, showing closer clustering and low diversity. Mean values and summations are previously derived, so only direct calculations are presented—starting with $median(x_j)$ computation for each design variable:

Exploration phase:

$$median(x_1) = 4.9961 \quad median(x_2) = 3.6837 \quad median(x_3) = 5.0100$$

Exploitation phase:

$$median(x_1) = 3.5723 \quad median(x_2) = 3.2483 \quad median(x_3) = 3.5076$$

Once the median values are assessed, the next step is to calculate the diversity as the average absolute deviation from the median for each dimension d_j as in Equation 7.12, obtaining the following results:

Exploration phase:

$$d_{j_1} = 1.7898 \quad d_{j_2} = 2.2738 \quad d_{j_3} = 2.2404$$

Exploitation phase:

$$d_{j_1} = 0.2070 \quad d_{j_2} = 0.3428 \quad d_{j_3} = 0.2644$$

Then the overall dimension-wise population diversity d_{DWD} for each case is calculated by Equation 7.13 giving the following results:

Exploration phase:

$$d_{DWD} = 2.0813$$

Exploitation phase:

$$d_{DWD} = 0.2714$$

The dimension-wise diversity results are 2.0813 for the exploration phase and 0.2714 for the exploitation phase. These results indicate the average distance of the search agents from the median across all dimensions. In turn, it can be inferred that if d_{DWD} increases, the population is exploring more of the search space; in other words, exploration dominates. If d_{DWD} decreases over iterations, the population converges (exploitation dominates). The last step is to observe the current behavior of the population based on the $ExplorationRate$ and the $ExploitationRate$ computed with Equations 7.14 and 7.15, respectively, for each case. The maximum diversity d_{max} is typically observed during optimization. For this example, let us assume $d_{max} = 3.5$; in practice, using the actual maximum diversity observed during your optimization is necessary. The results are:

Exploration phase:

$$ExplorationRate \approx 59.47$$

$$ExploitationRate \approx 40.53$$

Exploitation phase:

$$ExplorationRate \approx 7.75$$

$$ExploitationRate \approx 92.25$$

For the exploration phase test case, the results for the $ExplorationRate = 59.47$ and the $ExploitationRate = 40.53$ depict that the current state of the population is actively exploring the search space and exploiting and refining known adequate solutions. The results show a slight bias towards exploration over exploitation. This balance is generally desirable in many optimization problems since it allows the algorithm to explore new regions of the search space to discover potentially better solutions and exploit known good solutions.

The exploitation phase test case obtained the results of $ExplorationRate = 7.75$ and $ExploitationRate = 92.25$. The population is slightly exploring but also converging on the best solution. However, premature convergence may occur, trapping the algorithm in a local optimum. A low exploration rate signals low diversity, meaning similar search agents. While acceptable in later optimization stages, it is problematic early on, as limited exploration reduces the chances of finding better solutions.

Figure 7.8 presents the distributions of the two population sets taken from Tables 7.5 and 7.6. As seen in the figure, the circles represent the distribution of search agents for the exploration phase test case, and the triangles the search agents in the exploitation phase. The $median$ particle for each test case is depicted by a star marker, with yellow for the $median$ of the exploration phase and green for the $median$ of the exploitation phase. The dashed line represents the distance of each particle to their respective median. In turn, the computed dimension-wise diversity values are shown at the top of the graph.

Understanding and leveraging the dimension-wise diversity metric is essential for balancing exploration and exploitation in metaheuristic algorithms, ultimately enhancing

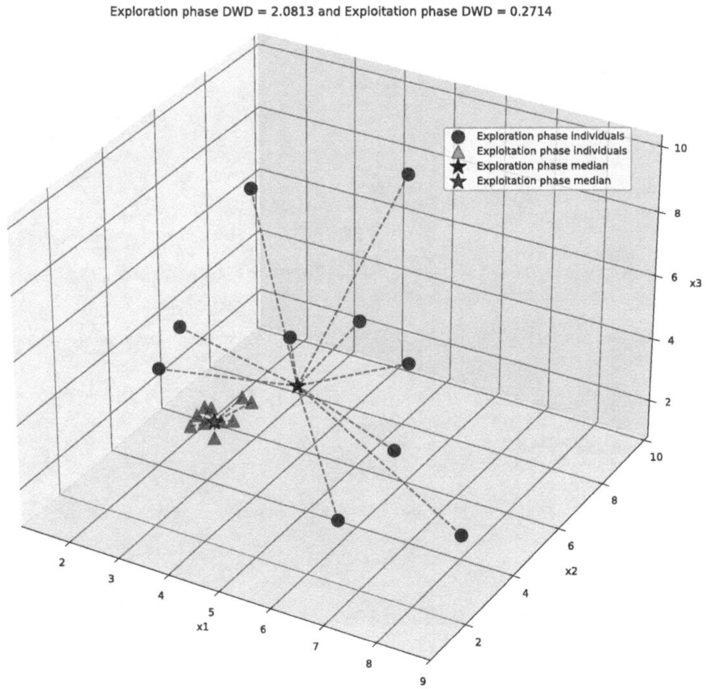

Figure 7.8: Differences in the distribution of the population in the exploration and exploitation phase.

their performance in solving complex optimization problems. Using the median value in the balance response prevents inconsistencies by providing a stable reference point. Additionally, the influence of d_{max} on $ExplorationRate$ and $ExploitationRate$ ensures that the rates of exploration and exploitation are evaluated against the maximum diversity encountered, providing a robust measure of the algorithm's search dynamics. However, the dimension-wise diversity metric has a filtering effect due to the averaging of all dimensions, which can partially smoothen abrupt changes in diversity and may result in a slight bias toward exploitation. The Pseudocode and the MATLAB code for the DWD assessment are presented below.

7.5.1.1 Pseudocode

```
BEGIN Calculate Dimension-wise Diversity

# Define the problem
Define the population size N
Define the dimensionality of the search space S

# Initialize variables
Median_Vector = [0] * S  # Vector to store median values of each dimension
Diversity_Vector = [0] * S  # Vector to store diversity values for each
    dimension

# Compute median values for each dimension
FOR each dimension d in (1 to S)
    Sort the values of dimension d across all individuals in Population P
    IF N is odd THEN
        Median_Vector[d] = P[(N + 1) / 2][d]  # Middle value
    ELSE
        Median_Vector[d] = (P[N / 2][d] + P[(N / 2) + 1][d]) / 2  # Average of
    two middle values
    END IF
END FOR

# Compute dimension-wise diversity
FOR each dimension d in (1 to S)
    Sum_Abs_Dev = 0
    FOR each individual i in Population P (i = 1 to N)
        Sum_Abs_Dev = Sum_Abs_Dev + |P[i][d] - Median_Vector[d]|  # Absolute
    deviation from median
    END FOR
    Diversity_Vector[d] = Sum_Abs_Dev / N  # Average absolute deviation
END FOR

# Compute overall dimension-wise diversity (DWD)
DWD = 0
FOR each dimension d in (1 to S)
    DWD = DWD + Diversity_Vector[d]
END FOR
DWD = DWD / S  # Average across all dimensions

# Output the result
RETURN DWD

END Calculate Dimension-wise Diversity
```

Pseudocode explanation

- Input Parameters: The function defines the population size (N) and the dimension of the search space (S). A vector Median_Vector of size S is initialized to store the median values for each dimension. Another vector, Diversity_Vector, of size S is initialized to store the diversity values for each dimension.
- Calculating the Diversity: The first loop iterates over all dimensions (d) to compute the median value of each dimension across the population. If N is odd, the median is selected as the middle value in the sorted list. If N is even, the median is calculated as the average of the two middle values. In the second step, another loop iterates over each dimension (d) to compute the average absolute deviation from the median. For each individual (i) in the population, the absolute difference between the individual's value and the median is computed and summed. The diversity for each dimension is then obtained by dividing this sum by N.
- Result: Finally, the function returns the Dimension-wise Diversity value as a measure of diversity in the evolutionary algorithm's population.

7.5.1.2 MATLAB Code

```
function DWD = DimensionWiseDiversity(population)
    % DimensionWiseDiversity - Computes the dimension-wise diversity (DWD)
    % of a population based on median absolute deviation.
    % population - Matrix (NxD) where N is the number of individuals
    %              and D is the dimensionality of the search space.

    % Population Size and Dimensionality
    [numParticles, numDimensions] = size(population);

    % Initialize vectors
    dimensionalMedians = zeros(1, numDimensions);
    averageDistance = zeros(1, numDimensions);

    % Compute median for each dimension
    for j = 1:numDimensions
        dimensionalMedians(j) = median(population(:, j)); % Using MATLAB's
    median function
    end

    % Compute absolute deviations from median and their averages
    for j = 1:numDimensions
        averageDistance(j) = sum(abs(population(:, j) - dimensionalMedians(j)))
    / numParticles;
    end

    % Compute overall dimension-wise diversity (DWD)
    dimensionalDiversity = sum(averageDistance);
    DWD = dimensionalDiversity / numDimensions; % Normalize by number of
    dimensions

    % Display result
    fprintf('Computed Dimension-Wise Diversity (DWD): %.4f\n', DWD);
end
```

Example input for testing the code in MATLAB

```
% Define a population matrix of 10 individuals in a 7-dimensional space
population = [
    1.2, 9.8, 4.5, 7.1, 3.3, 6.9, 2.1;
    2.5, 8.3, 5.9, 9.2, 4.1, 3.7, 6.5;
    3.9, 7.1, 8.2, 2.3, 5.5, 9.6, 4.4;
    6.4, 2.8, 9.5, 5.7, 7.6, 4.2, 8.9;
    9.1, 3.4, 2.7, 6.9, 8.1, 5.8, 1.9;
    4.7, 6.2, 1.5, 8.6, 9.9, 7.3, 3.2;
    7.5, 5.1, 6.8, 4.9, 2.2, 1.8, 9.7;
    8.3, 1.6, 7.4, 3.5, 6.7, 2.9, 5.4;
    5.6, 4.3, 3.9, 1.7, 9.4, 8.5, 7.8;
    10.0, 2.1, 5.2, 9.8, 1.3, 6.1, 4.0
];

% Call the function
DWD = DimensionWiseDiversity(population);
```

References

[1] V. Osuna-Enciso, E. Cuevas, and B. M. Castañeda, "A diversity metric for population-based metaheuristic algorithms," *Information Sciences*, vol. 586, pp. 192–208, 2022. [Online]. Available: https://www.sciencedirect.com/science/article/pii/S0020025521011956

[2] K. Hussain, M. Najib, S. Cheng, Y. Shi, and R. Naseem, "Artificial bee colony algorithm: A component-wise analysis using diversity measurement," *Journal of King Saud University - Computer and Information Sciences*, vol. 32, no. 7, pp. 794–808, 2020. [Online]. Available: https://www.sciencedirect.com/science/article/pii/S1319157818302039

[3] S. Cheng, Y. Shi, Q. Qin, Q. Zhang, and R. Bai, "Population diversity maintenance in brain storm optimization algorithm," *Journal of Artificial Intelligence and Soft Computing Research (JAISCR)*, vol. 4, pp. 83–97, 10 2014.

[4] D. Karaboga *et al.*, "An idea based on honey bee swarm for numerical optimization," Technical report-tr06, Erciyes university, engineering faculty, computer . . . , Tech. Rep., 2005.

[5] D. Tian, Q. Xu, X. Yao, G. Zhang, Y. Li, and C. Xu, "Diversity-guided particle swarm optimization with multi-level learning strategy," *Swarm and Evolutionary Computation*, vol. 86, p. 101533, 2024.

[6] R. Poláková, J. Tvrdík, and P. Bujok, "Adaptation of population size according to current population diversity in differential evolution," in *2017 IEEE Symposium Series on Computational Intelligence (SSCI)*, 2017, pp. 1–8.

[7] R. Poláková, J. Tvrdík, and P. Bujok, "Differential evolution with adaptive mechanism of population size according to current population diversity," *Swarm and Evolutionary Computation*, vol. 50, p. 100519, 2019.

[8] M. Yang, C. Li, Z. Cai, and J. Guan, "Differential evolution with auto-enhanced population diversity," *IEEE Transactions on Cybernetics*, vol. 45, no. 2, pp. 302–315, 2015.

[9] M. Wineberg and F. Oppacher, "The underlying similarity of diversity measures used in evolutionary computation," in *Genetic and Evolutionary Computation — GECCO 2003*. Berlin, Heidelberg: Springer Berlin Heidelberg, 2003, pp. 1493–1504.

[10] R. Fabbri, L. D. F. Costa, J. C. Torelli, and O. M. Bruno, "2d euclidean distance transform algorithms: A comparative survey," *ACM Computing Surveys (CSUR)*, vol. 40, no. 1, pp. 1–44, 2008.

[11] S. J. Louis and G. J. Rawlins, "Syntactic analysis of convergence in genetic algorithms," in *Foundations of Genetic Algorithms*. Elsevier, 1993, vol. 2, pp. 141–151.

[12] A. Wehrl, "General properties of entropy," *Reviews of Modern Physics*, vol. 50, no. 2, p. 221, 1978.

[13] I. A. Mageed and Q. Zhang, "An introductory survey of entropy applications to information theory, queuing theory, engineering, computer science, and statistical mechanics," in *2022 27th International Conference on Automation and Computing (ICAC)*. IEEE, 2022, pp. 1–6.

[14] M. Ribeiro, T. Henriques, L. Castro, A. Souto, L. Antunes, C. Costa-Santos, and A. Teixeira, "The entropy universe," *Entropy*, vol. 23, no. 2, p. 222, 2021.

[15] C. E. Shannon, "A mathematical theory of communication," *The Bell System Technical Journal*, vol. 27, no. 4, pp. 623–656, 1948.

[16] L. Jost, "Entropy and diversity," *Oikos*, vol. 113, no. 2, pp. 363–375, 2006.

[17] A. E. Eiben and S. K. Smit, "Evolutionary algorithm parameters and methods to tune them," in *Autonomous Search*. Springer, 2012, pp. 15–36.

[18] S. Ghosh, S. Das, S. Roy, S. M. Islam, and P. N. Suganthan, "A differential covariance matrix adaptation evolutionary algorithm for real parameter optimization," *Information Sciences*, vol. 182, no. 1, pp. 199–219, 2012.

[19] D. Sudholt, "The benefits of population diversity in evolutionary algorithms: a survey of rigorous runtime analyses," *Theory of Evolutionary Computation: Recent Developments in Discrete Optimization*, pp. 359–404, 2019.

[20] E. K. Burke, S. Gustafson, and G. Kendall, "Diversity in genetic programming: An analysis of measures and correlation with fitness," *IEEE Transactions on Evolutionary Computation*, vol. 8, no. 1, pp. 47–62, 2004.

[21] S. Cheng, Y. Shi, Q. Qin, Q. Zhang, and R. Bai, "Population diversity maintenance in brain storm optimization algorithm," *Journal of Artificial Intelligence and Soft Computing Research*, vol. 4, no. 2, pp. 83–97, 2014.

Chapter 8

Implementation of Diversity Indicators in DE and PSO

This chapter explores improving Differential Evolution (DE) [1] and Particle Swarm Optimization (PSO) [2] by integrating dimension-wise diversity. Maintaining diversity prevents premature convergence, ensuring a balance between exploration (searching new regions) and exploitation (refining optimal solutions).

The chapter is divided into two main sections. Each section focuses on implementing these diversity indicators in the respective algorithms, DE and PSO. Through detailed explanations and exercises, you will learn how to integrate diversity metrics and analyze their impact on the algorithms' ability to solve complex optimization problems.

8.1 Analysis of Diversity

8.1.1 *Implementing Dimension-wise Diversity Metric on Differential Evolution Algorithm*

The following MATLAB® code implements the Differential Evolution (DE) [1] algorithm with the addition of a diversity metric [3]. The purpose is twofold:

- Solve Optimization Problems: The code optimizes a problem defined by a fitness function (FName). It uses a population of candidate solutions that evolve iteratively to find the best solution within specified bounds.

DOI: 10.1201/9781032723563-8

- Monitor Diversity: The code calculates and tracks a dimension-wise diversity metric during the optimization process. This metric measures the spread of the population in the search space and helps to evaluate how well the algorithm balances exploration and exploitation.

Understanding and implementing diversity metrics are essential for improving meta-heuristic algorithms and ensuring they do not converge prematurely to suboptimal solutions.

8.1.2 *MATLAB Code*

```
function [best_individual, best_fitness] = DifferentialEvolution(f, NP, G, F, CR
    , lower_bound, upper_bound)
    % f: Objective function handle
    % NP: Number of individuals in the population
    % G: Maximum number of generations
    % F: Differential weight
    % CR: Crossover rate
    % lower_bound: Lower bound of the search space
    % upper_bound: Upper bound of the search space

    % Initialize population
    dimensions = length(lower_bound);
    population = lower_bound + (upper_bound - lower_bound) .* rand(NP,
     dimensions);
    fitness = arrayfun(f, population);

    % Initialize diversity
    highestDiversity = 0;
    dimensionalMedians = zeros(1,dimensions);
    for j=1:dimensions
        dimensionalPositions = zeros(NP,1);
        for i=1:NP
            dimensionalPositions(i) = population(i, j);
        end
        dimensionalMedians(j)=median(dimensionalPositions);
    end

    averageDistance = zeros(1,dimensions);
    for j=1:dimensions
        distancetoMedian = 0;
        for i=1:NP
            distancetoMedian = distancetoMedian + abs(population(i, j) -
     dimensionalMedians(j));
        end
        averageDistance(j)=distancetoMedian/NP;
    end

    dimensionalDiversity=sum(averageDistance(:));
    averageDiversity = dimensionalDiversity/dimensions;
    if averageDiversity > highestDiversity
        highestDiversity = averageDiversity;
    end

    % Main loop
    for generation = 1:G
        for i = 1:NP
            % Mutation: Create a mutant vector
            indices = randperm(NP, 3);
            while any(indices == i)
                indices = randperm(NP, 3);
            end
            r1 = indices(1);
```

```
            r2 = indices(2);
            r3 = indices(3);
            mutant_vector = population(r1, :) + F * (population(r2, :) -
    population(r3, :));

            % Ensure the mutant vector is within the bounds
            mutant_vector = max(mutant_vector, lower_bound);
            mutant_vector = min(mutant_vector, upper_bound);

            % Crossover: Create a trial vector
            trial_vector = population(i, :);
            j_rand = randi(dimensions);
            for j = 1:dimensions
                if rand <= CR || j == j_rand
                    trial_vector(j) = mutant_vector(j);
                end
            end

            % Selection: Select the better between the trial vector and the
    target vector
            trial_fitness = f(trial_vector);
            if trial_fitness < fitness(i)
                population(i, :) = trial_vector;
                fitness(i) = trial_fitness;
            end
        end

        % Update diversity
        dimensionalMedians = zeros(1,dimensions);
        for j=1:dimensions
            dimensionalPositions = zeros(NP,1);
            for i=1:NP
                dimensionalPositions(i) = population(i, j);
            end
            dimensionalMedians(j)=median(dimensionalPositions);
        end

        averageDistance = zeros(1,dimensions);
        for j=1:dimensions
            distancetoMedian = 0;
            for i=1:NP
                distancetoMedian = distancetoMedian + abs(population(i, j) -
    dimensionalMedians(j));
            end
            averageDistance(j)=distancetoMedian/NP;
        end

        dimensionalDiversity=sum(averageDistance(:));
        averageDiversity = dimensionalDiversity/dimensions;
        if averageDiversity > highestDiversity
            highestDiversity = averageDiversity;
        end
        percentualDiversity = averageDiversity*100/highestDiversity;

    end

    % Output the best solution found
    [best_fitness, idx] = min(fitness);
    best_individual = population(idx, :);
end
```

Now, let us break down the code into its key sections to understand its structure and functionality.

8.1.3 Problem Definition and Algorithm Parameters

```
function [best_individual, best_fitness] = DifferentialEvolution(f, NP, G, F, CR
    , lower_bound, upper_bound)
```

The algorithm receives the optimization problem to solve as input parameters. The following list contains an explanation of each parameter.

- f: Objective function handle for the optimization problem.
- G: Maximum number of generations (iterations) allowed.
- $lower\_bound$: Minimum values permissible for each of the decision variables.
- $upper\_bound$: Maximum values permissible for each of the decision variables.

On the other hand, the behavior of the Differential Evolution algorithm is controlled by the following set of input parameters.

- NP: Number of candidate solutions (population size).
- F: Scaling factor for mutation.
- CR: Probability of crossover.

These parameters significantly influence the algorithm performance and must be set correctly for the best results.

8.1.4 Algorithm Initialization

```
% Initialize population
dimensions = length(lower_bound);
population = lower_bound + (upper_bound - lower_bound) .* rand(NP, dimensions);
fitness = arrayfun(f, population);
```

The population of candidate solutions is initialized. This process is accomplished with the following steps:

- Initialize the $dimensions$ variable that contains the number of decision variables required by the optimization problem.
- Randomly generate a $population$ within the specified lower and upper bounds [4].
- Evaluate the fitness of each individual using MATLAB's $arrayfun$ function.

8.1.5 Diversity Metric Initialization

```
% Initialize diversity
highestDiversity = 0;
dimensionalMedians = zeros(1,dimensions);
for j=1:dimensions
    dimensionalPositions = zeros(NP,1);
    for i=1:NP
        dimensionalPositions(i) = population(i, j);
    end
```

```
    dimensionalMedians(j)=median(dimensionalPositions);
end

averageDistance = zeros(1,dimensions);
for j=1:dimensions
    distancetoMedian = 0;
    for i=1:NP
        distancetoMedian = distancetoMedian + abs(population(i, j) -
     dimensionalMedians(j));
    end
    averageDistance(j)=distancetoMedian/NP;
end

dimensionalDiversity=sum(averageDistance(:));
averageDiversity = dimensionalDiversity/dimensions;
if averageDiversity > highestDiversity
    highestDiversity = averageDiversity;
end
```

A diversity metric is computed to measure the spread of solutions in each dimension. This is done through the following set of steps.

- Calculate the median of each dimension. This is saved to the $dimensionalMedians$ variable.
- Calculate the average deviation ($averageDistance$) of each population member from these medians.
- Aggregate this deviation across dimensions to obtain the overall diversity in each dimension ($dimensionalDiversity$).
- Dimensional diversity is averaged to obtain the complete population average diversity ($averageDiversity$).

8.1.6 *Differential Evolution Main Loop*

```
for i = 1:NP
    % Mutation: Create a mutant vector
    indices = randperm(NP, 3);
    while any(indices == i)
        indices = randperm(NP, 3);
    end
    r1 = indices(1);
    r2 = indices(2);
    r3 = indices(3);
    mutant_vector = population(r1, :) + F * (population(r2, :) - population(r3,
     :));

    % Ensure the mutant vector is within the bounds
    mutant_vector = max(mutant_vector, lower_bound);
    mutant_vector = min(mutant_vector, upper_bound);

    % Crossover: Create a trial vector
    trial_vector = population(i, :);
    j_rand = randi(dimensions);
    for j = 1:dimensions
        if rand <= CR || j == j_rand
            trial_vector(j) = mutant_vector(j);
        end
    end
```

```
    % Selection: Select the better between the trial vector and the target
    vector
    trial_fitness = f(trial_vector);
    if trial_fitness < fitness(i)
        population(i, :) = trial_vector;
        fitness(i) = trial_fitness;
    end
end
```

This is the heart of the algorithm, where the population evolves.

8.1.7 *Mutation*

```
% Mutation: Create a mutant vector
indices = randperm(NP, 3);
while any(indices == i)
    indices = randperm(NP, 3);
end
r1 = indices(1);
r2 = indices(2);
r3 = indices(3);
mutant_vector = population(r1, :) + F * (population(r2, :) - population(r3, :));

% Ensure the mutant vector is within the bounds
mutant_vector = max(mutant_vector, lower_bound);
mutant_vector = min(mutant_vector, upper_bound);
```

The first significant movement operator is the Mutation. Here, the algorithm creates a *mutant* vector ($mutant\_vector$) by combining three randomly selected individuals ($population(r1,:), population(r2,:) and population(r3,:)$). The scaling factor F controls the influence of the difference between individuals two and three.

8.1.8 *Crossover*

```
% Crossover: Create a trial vector
trial_vector = population(i, :);
j_rand = randi(dimensions);
for j = 1:dimensions
    if rand <= CR || j == j_rand
        trial_vector(j) = mutant_vector(j);
    end
end
```

The crossover operator combines the *mutant* vector with the current solution to generate a new trial vector ($trial\_vector$). The process begins by copying the current solution into the trial vector. Then, for each dimension, a crossover probability (CR) is applied to decide whether the corresponding value is replaced by the value from the mutant vector. To ensure a minimum amount of crossover between the solutions, a dimension (j_rand) is randomly selected to always be replaced. This probabilistic mechanism controls the degree of mixing between the two vectors.

8.1.9 *Selection*

```
% Selection: Select the better between the trial vector and the target vector
trial_fitness = f(trial_vector);
if trial_fitness < fitness(i)
    population(i, :) = trial_vector;
    fitness(i) = trial_fitness;
end
```

The new solution is retained only if it improves the fitness of the current solution. If the $(trial\_vector)$ is no better than the current solution, it is discarded.

8.1.10 *Diversity Update*

```
% Update diversity
dimensionalMedians = zeros(1,dimensions);
for j=1:dimensions
    dimensionalPositions = zeros(NP,1);
    for i=1:NP
        dimensionalPositions(i) = population(i, j);
    end
    dimensionalMedians(j)=median(dimensionalPositions);
end

averageDistance = zeros(1,dimensions);
for j=1:dimensions
    distancetoMedian = 0;
    for i=1:NP
        distancetoMedian = distancetoMedian + abs(population(i, j) -
     dimensionalMedians(j));
    end
    averageDistance(j)=distancetoMedian/NP;
end

dimensionalDiversity=sum(averageDistance(:));
averageDiversity = dimensionalDiversity/dimensions;
if averageDiversity > highestDiversity
    highestDiversity = averageDiversity;
end
percentualDiversity = averageDiversity*100/highestDiversity;
```

At each iteration, the diversity metric is recalculated to track how the population spreads over time. This recalculation is accomplished following the same guidelines as in the diversity initialization process.

8.1.11 *Results and Output*

```
% Output the best solution found
[best_fitness, idx] = min(fitness);
best_individual = population(idx, :);
```

The code tracks and returns:

- $best\_individual$: The optimal solution which produced the best fitness after the optimization process.
- $best\_fitness$: The minimum cost produced by the best found solution.

8.2 Diversity Analysis in Particle Search Optimization

8.2.1 *Implementing Dimension-wise Diversity Metric on Particle Swarm Optimization Algorithm*

This MATLAB code implements the Particle Swarm Optimization (PSO) [2] algorithm with an added diversity metric [3]. It serves two main purposes:

- Optimization of Objective Functions: The code solves optimization problems by simulating the behavior of particles in a swarm. Each particle represents a potential solution, and the swarm collectively explores the search space.
- Tracking Population Diversity: A diversity metric is computed and monitored to assess the spread of particles in the search space. This metric provides insights into the algorithm's balance between exploration (searching broadly) and exploitation (refining known good solutions).

By including exercises to modify and analyze the diversity metric, the code encourages a deeper understanding of the relationship between diversity and optimization performance.

8.2.2 *MATLAB Code*

```
function [global_best_individual, global_best_fitness] =
    ParticleSwarmOptimization(f, NP, T, w, c1, c2, lower_bound, upper_bound)
  % f: Objective function handle
  % NP: Number of particles
  % T: Maximum number of iterations
  % w: Inertia weight
  % c1: Cognitive constant
  % c2: Social constant
  % lower_bound: Lower bound of the search space
  % upper_bound: Upper bound of the search space


  % Initialize particle positions and velocities
  dimensions = length(lower_bound);
  position = lower_bound + (upper_bound - lower_bound) .* rand(NP, dimensions)
  ;
  personal_best_position = position;
  velocity = zeros(NP, dimensions);
  fitness = arrayfun(f, position);

  % Initialize global best
  [global_best_fitness, idx] = min(fitness);
  global_best_individual = personal_best_position(idx, :);

  % Initialize diversity
  highestDiversity = 0;
  dimensionalMedians = zeros(1,dims);
  for j=1:dims
      dimensionalPositions = zeros(NP,1);
      for i=1:NP
          dimensionalPositions(i) = position(i, j);
      end
      dimensionalMedians(j)=median(dimensionalPositions);
  end

  averageDistance = zeros(1,dimensions);
  for j=1:dimensions
```

```
        distancetoMedian = 0;
        for i=1:NP
            distancetoMedian = distancetoMedian + abs(position(i, j) -
 dimensionalMedians(j));
        end
        averageDistance(j)=distancetoMedian/NP;
end

dimensionalDiversity=sum(averageDistance(:));
averageDiversity = dimensionalDiversity/dimensions;
if averageDiversity > highestDiversity
    highestDiversity = averageDiversity;
end

% Main loop
for t = 1:T
    for i = 1:NP
        % Update velocity
        r1 = rand(1, dimensions);
        r2 = rand(1, dimensions);
        velocity(i, :) = w * velocity(i, :) + ...
                        c1 * r1 .* (personal_best_position(i, :) - position
 (i, :)) + ...
                        c2 * r2 .* (global_best_individual - position(i, :)
 );

        % Update position
        position(i, :) = position(i, :) + velocity(i, :);

        % Ensure the position is within the bounds
        position(i, :) = max(position(i, :), lower_bound);
        position(i, :) = min(position(i, :), upper_bound);

        % Evaluate the new position
        current_fitness = f(position(i, :));

        % Update the particle's best known position
        if current_fitness < fitness(i)
            personal_best_position(i, :) = position(i, :);
            fitness(i) = current_fitness;
        end

        % Update the global best known position
        if current_fitness < global_best_fitness
            global_best_individual = position(i, :);
            global_best_fitness = current_fitness;
        end
    end

    % Update diversity
    highestDiversity = 0;
    dimensionalMedians = zeros(1,dims);
    for j=1:dims
        dimensionalPositions = zeros(NP,1);
        for i=1:NP
            dimensionalPositions(i) = position(i, j);
        end
        dimensionalMedians(j)=median(dimensionalPositions);
    end

    averageDistance = zeros(1,dimensions);
    for j=1:dimensions
        distancetoMedian = 0;
        for i=1:NP
            distancetoMedian = distancetoMedian + abs(position(i, j) -
 dimensionalMedians(j));
        end
```

```
            averageDistance(j)=distancetoMedian/NP;
        end

        dimensionalDiversity=sum(averageDistance(:));
        averageDiversity = dimensionalDiversity/dimensions;
        if averageDiversity > highestDiversity
            highestDiversity = averageDiversity;
        end
        percentualDiversity = averageDiversity*100/highestDiversity;

        % Optionally, adjust parameters such as inertia weight
        % w = UpdateInertiaWeight(w, t, T);
    end

    % Output the best solution found
    [global_best_fitness, idx] = min(fitness);
    global_best_individual = position(idx, :);
end
```

Now, let us break down the code into its key sections to understand its structure and functionality.

8.2.3 *Problem Definition*

```
function [global_best_individual, global_best_fitness] =
    ParticleSwarmOptimization(f, NP, T, w, c1, c2, lower_bound, upper_bound)
```

The algorithm receives the optimization function as input parameters. The following list contains an explanation of each parameter.

- f: Objective function handle for the optimization problem.
- T: Maximum number of iterations allowed.
- *lower_bound*: Minimum values permissible for each of the decision variables.
- *upper_bound*: Maximum values permissible for each of the decision variables.

On the other hand, the following set of input parameters governs how particles move and adapt in the PSO algorithm:

- w: Inertia Weight. Controls the influence of the previous velocity of a particle.
- $c1$, $c2$: Learning coefficients. Balance between personal (individual particle) and global (swarm) knowledge.

These parameters ensure controlled and efficient search behavior.

8.2.4 *Algorithm Initialization*

```
% Initialize particle positions and velocities
dimensions = length(lower_bound);
position = lower_bound + (upper_bound - lower_bound) .* rand(NP, dimensions);
personal_best_position = position;
velocity = zeros(NP, dimensions);
fitness = arrayfun(f, position);

% Initialize global best
```

```
[global_best_fitness, idx] = min(fitness);
global_best_individual = personal_best_position(idx, :);
```

The population of candidate solutions is initialized. This process is accomplished with the following steps:

- Initialize the $dimensions$ variable that contains the number of decision variables required by the optimization problem.
- Randomly generate a $position$ within the specified lower and upper bounds [4].
- Initialize the $velocity$ variable to zero for all search agents.
- Evaluate the fitness of each individual using MATLAB's $arrayfun$ function.
- Select the best position in the entire swarm and save to the variables $global\_best\_individual$ and $global\_best\_fitness$.

8.2.5 *Diversity Metric Initialization*

```
% Initialize diversity
highestDiversity = 0;
dimensionalMedians = zeros(1,dims);
for j=1:dims
    dimensionalPositions = zeros(NP,1);
    for i=1:NP
        dimensionalPositions(i) = position(i, j);
    end
    dimensionalMedians(j)=median(dimensionalPositions);
end

averageDistance = zeros(1,dimensions);
for j=1:dimensions
    distancetoMedian = 0;
    for i=1:NP
        distancetoMedian = distancetoMedian + abs(position(i, j) -
     dimensionalMedians(j));
    end
    averageDistance(j)=distancetoMedian/NP;
end

dimensionalDiversity=sum(averageDistance(:));
averageDiversity = dimensionalDiversity/dimensions;
if averageDiversity > highestDiversity
    highestDiversity = averageDiversity;
end
```

A diversity metric is computed to measure the spread of solutions in each dimension. This is done through the following set of steps.

- Calculate the median of each dimension. This is saved to the $dimensionalMedians$ variable.
- Calculate the average deviation ($averageDistance$) of each population member from these medians.

- Aggregate this deviation across dimensions to obtain the overall diversity in each dimension ($dimensionalDiversity$).
- Dimensional diversity is averaged to obtain the complete population average diversity ($averageDiversity$).

8.2.6 *PSO Main Loop*

```
% Main loop
for t = 1:T
    for i = 1:NP
        % Update velocity
        r1 = rand(1, dimensions);
        r2 = rand(1, dimensions);
        velocity(i, :) = w * velocity(i, :) + ...
                         c1 * r1 .* (personal_best_position(i, :) - position(i,
:)) + ...
                         c2 * r2 .* (global_best_individual - position(i, :));

        % Update position
        position(i, :) = position(i, :) + velocity(i, :);

        % Ensure the position is within the bounds
        position(i, :) = max(position(i, :), lower_bound);
        position(i, :) = min(position(i, :), upper_bound);

        % Evaluate the new position
        current_fitness = f(position(i, :));

        % Update the particle's best known position
        if current_fitness < fitness(i)
            personal_best_position(i, :) = position(i, :);
            fitness(i) = current_fitness;
        end

        % Update the global best known position
        if current_fitness < global_best_fitness
            global_best_individual = position(i, :);
            global_best_fitness = current_fitness;
        end
    end

    % Update diversity
    highestDiversity = 0;
    dimensionalMedians = zeros(1,dims);
    for j=1:dims
        dimensionalPositions = zeros(NP,1);
        for i=1:NP
            dimensionalPositions(i) = position(i, j);
        end
        dimensionalMedians(j)=median(dimensionalPositions);
    end

    averageDistance = zeros(1,dimensions);
    for j=1:dimensions
        distancetoMedian = 0;
        for i=1:NP
            distancetoMedian = distancetoMedian + abs(position(i, j) -
 dimensionalMedians(j));
        end
        averageDistance(j)=distancetoMedian/NP;
    end

    dimensionalDiversity=sum(averageDistance(:));
```

```
    averageDiversity = dimensionalDiversity/dimensions;
    if averageDiversity > highestDiversity
        highestDiversity = averageDiversity;
    end
    percentualDiversity = averageDiversity*100/highestDiversity;

    % Optionally, adjust parameters such as inertia weight
    % w = UpdateInertiaWeight(w, t, T);
end
```

This loop runs for the specified number of iterations (T), during which particles adjust their positions and velocities.

8.2.7 *Velocity Update*

```
% Update velocity
r1 = rand(1, dimensions);
r2 = rand(1, dimensions);
velocity(i, :) = w * velocity(i, :) + ...
                 c1 * r1 .* (personal_best_position(i, :) - position(i, :)) +
    ...
                 c2 * r2 .* (global_best_individual - position(i, :));
```

The velocity update combines four components: inertia weight w, current velocity $velocity$, attraction to personal best $c1$, and attraction to the global best $c2$.

8.2.8 *Position Update*

```
% Update position
position(i, :) = position(i, :) + velocity(i, :);
```

The particle's $position$ changes based on its updated $velocity$.

8.2.9 *Boundary Handling*

```
% Ensure the position is within the bounds
position(i, :) = max(position(i, :), lower_bound);
position(i, :) = min(position(i, :), upper_bound);
```

Particle's $position$ and $velocity$ are constrained within the predefined limits ($lower\_bound$, $upper\_bound$).

8.2.10 *Evaluation*

```
% Evaluate the new position
current_fitness = f(position(i, :));

% Update the particle's best known position
if current_fitness < fitness(i)
    personal_best_position(i, :) = position(i, :);
    fitness(i) = current_fitness;
end

% Update the global best known position
if current_fitness < global_best_fitness
    global_best_individual = position(i, :);
    global_best_fitness = current_fitness;
end
```

The fitness of each particle according to their position (*current_fitness*) is evaluated. The best personal and global solutions are updated if the particle achieves better fitness (*personal_best_position*, *global_best_individual*).

8.2.11 *Update Diversity*

```
% Update diversity
highestDiversity = 0;
dimensionalMedians = zeros(1,dims);
for j=1:dims
    dimensionalPositions = zeros(NP,1);
    for i=1:NP
        dimensionalPositions(i) = position(i, j);
    end
    dimensionalMedians(j)=median(dimensionalPositions);
end

averageDistance = zeros(1,dimensions);
for j=1:dimensions
    distancetoMedian = 0;
    for i=1:NP
        distancetoMedian = distancetoMedian + abs(position(i, j) -
     dimensionalMedians(j));
    end
    averageDistance(j)=distancetoMedian/NP;
end

dimensionalDiversity=sum(averageDistance(:));
averageDiversity = dimensionalDiversity/dimensions;
if averageDiversity > highestDiversity
    highestDiversity = averageDiversity;
end
percentualDiversity = averageDiversity*100/highestDiversity;
```

After each iteration, the diversity is recalculated to monitor the spread of particles in the search space. This helps analyze the behavior of the swarm over time.

8.2.12 *Results and Output*

```
% Output the best solution found
[global_best_fitness, idx] = min(fitness);
global_best_individual = position(idx, :);
```

The code tracks and returns:

- *global_best_individual*: The optimal solution which produced the best fitness from the entire swarm.
- *global_best_fitness*: The minimum cost produced by the optimal solution.

8.3 Exercises for the Reader

To help readers internalize the concepts and gain hands-on experience, here are several exercises related to implementing and analyzing diversity metrics in the Differential Evolution algorithm and the Particle Swarm Optimization algorithm.

8.3.1 *Analyze the Diversity Metric*

Objective: Understand the diversity metric and its role in PSO.

Locate the section where the diversity metric (div) is calculated. Explain in your own words how it measures the spread of particles in the search space. Plot the diversity metric ($percentualDiversity$) across iterations.

Answer the following questions: How does diversity change as the algorithm progresses? What is the relationship between diversity and convergence of the swarm?

8.3.2 *Modify the Diversity Metric*

Objective: Experiment with alternative ways of measuring diversity.

Replace the median-based diversity metric with one of the following: Standard Deviation-Based Diversity: Use the standard deviation of the positions of the particles in each dimension, instead of the median.

Pairwise Distance Diversity: Calculate the average Euclidean distance between all pairs of particles. Add your new diversity metric to the code and compare it with the original implementation.

Discuss: How does your metric affect the algorithm's behavior and convergence? Which metric provides more meaningful insights for your problem?

8.3.3 *Experiment with Crossover Probability (CR) in DE Algorithm*

Objective: Analyze how parameters influence diversity and population dynamics.

The crossover probability (CR) influences how much information is shared between solutions.

Experiment with different values: Set $CR = 0.4$, $CR = 0.8$, and $CR = 1.0$. Track the diversity metric and best fitness for each case.

Questions to Answer: Does a lower or higher CR result in better exploration? How does CR affect diversity trends and convergence speed?

8.3.4 *Balancing Exploration and Exploitation on PSO Algorithm*

Objective: Analyze how parameters influence diversity and swarm dynamics.

Modify the PSO parameters (w, $c1$, $c2$) to create three scenarios: High exploration (e.g., larger w and lower $c1$, $c2$ values). High exploitation (e.g., smaller w and higher $c1$, $c2$ values). Balanced (moderate values for all parameters). Plot the diversity metric and global best fitness for each scenario.

Discuss: How do parameter settings impact diversity and convergence? Which scenario performs best for a multimodal function?

8.3.5 *Introduce a Diversity-Based Termination Criterion*

Objective: Explore the use of diversity as a control mechanism.

Modify the stopping condition of the DE and the PSO algorithms to include diversity.

For example: Stop the algorithm if diversity ($percentualDiversity$) falls below a certain threshold, e.g., $percentualDiversity < 10$.

Tasks: Implement this termination criterion in the code. Run the algorithm with and without the new criterion on the same problem.

Questions to Answer: Does the new criterion improve computational efficiency (fewer iterations)? Does it lead to better or worse solutions?

References

[1] R. Storn and K. Price, "Differential evolution – a simple and efficient heuristic for global optimization over continuous spaces," *J. of Global Optimization*, vol. 11, no. 4, p. 341–359, dec 1997. [Online]. Available: https://doi.org/10.1023/A:1008202821328

[2] J. Kennedy and R. Eberhart, "Particle swarm optimization," in *Proceedings of ICNN'95 - International Conference on Neural Networks*, vol. 4, 1995, pp. 1942–1948 vol.4.

[3] S. Cheng, Y. Shi, Q. Qin, Q. Zhang, and R. Bai, "Population diversity maintenance in brain storm optimization algorithm," *Journal of Artificial Intelligence and Soft Computing Research*, vol. 4, no. 2, pp. 83–97, 2014.

[4] W. D. Kelton, "Random initialization methods in simulation," *IIE Transactions*, vol. 21, no. 4, pp. 355–367, 1989.

Chapter 9

Pros and Cons of the Use of Different Initializations and Diversity Indicators

This chapter presents a brief discussion of the pros and cons of the employment of initialization and diversity indicators in evolutionary algorithms. The chapter is divided into two main sections. First, the advantages and disadvantages of initialization methodologies are depicted. The second section will focus on the pros and cons of the diversity metrics.

9.1 Initialization Methodologies

Initialization is crucial for an algorithm's effectiveness and performance, as seen in Chapter 4. It influences the algorithm's exploration, convergence speed, and solution quality. As the first step, it defines parameters and generates initial solutions. This chapter examines various initialization techniques, assessing their advantages and drawbacks.

Initialization Importance

Effective initialization enhances exploration and exploitation, improving convergence speed and solution quality. Selecting the right technique prevents delays and avoids premature convergence into suboptimal solutions. Awareness of these factors optimizes algorithm performance for better results.

DOI: 10.1201/9781032723563-9

Types of Initialization

Several initialization techniques cater to different problem characteristics and algorithmic requirements. Some of the most prominent categories include:

- **Random Initialization:** This technique generates solutions by randomly sampling values for each decision variable within the defined search space bounds. It is the simplest and most common method, requiring minimal computational effort. However, its random nature can lead to solutions far from the optimal region, resulting in slow convergence.
- **Heuristic Initialization:** This method leverages domain-specific knowledge to generate initial solutions that are closer to the optimal region. It utilizes heuristics or rules derived from experience or expert knowledge to guide the initial population generation. Heuristic initialization can accelerate convergence by starting with higher-quality solutions. However, it can limit diversity and may bias the search towards specific regions of the search space, potentially missing other promising areas.
- **Knowledge-Based Initialization:** This approach utilizes prior knowledge or data about the problem to create the initial population. It can involve using historical data, expert opinions, or insights derived from similar problems. Knowledge-based initialization can significantly improve the quality of initial solutions and convergence speed. However, it depends heavily on the quality and relevance of the available knowledge. Outdated or inaccurate knowledge can lead to biased or suboptimal solutions.
- **Deterministic Initialization:** Deterministic initialization generates solutions following a predefined pattern or rule, ensuring a structured and predictable distribution across the search space. This approach includes methods like grid-based initialization and low discrepancy sequences like Latin Hypercube Sampling and Sobol sequences. Deterministic methods promote consistent and repeatable experiments. They excel in structured problem domains by ensuring even search space coverage. However, they often lack diversity, leading to potential stagnation and difficulty handling high-dimensional problems.
- **Oversampling Initialization:** Oversampling involves generating a significantly larger initial population than required and then selecting a subset based on certain criteria, such as fitness or diversity. This method increases the chance of finding high-quality solutions in the initial population and can enhance diversity. Nevertheless, it requires increased computational effort to evaluate the larger set of initial solutions.
- **Hybrid Initialization:** Hybrid initialization combines multiple techniques to leverage their strengths and mitigate their weaknesses. For instance, a hybrid approach might incorporate random initialization for exploration and heuristic initialization for exploitation, or it could use deterministic initialization for structured coverage and oversampling for enhanced diversity.

Hybrid initialization offers flexibility and adaptability to various problem characteristics. However, it can be more complex to implement and requires careful parameter tuning to balance the contributions of different initialization techniques.

9.1.1 *Pros and Cons of Initialization*

Each initialization methodology has its own set of advantages and disadvantages. Understanding these trade-offs is crucial for selecting the appropriate technique for a given optimization problem. Table 9.1 depicts the pros and cons of initialization strategies.

Table 9.1: Advantages and disadvantages of the most commonly used initialization techniques.

Initialization Technique	Advantages	Disadvantages
Random [1]	Simple to implement. Promotes wide exploration.	Slow convergence. Initial solutions may be far from optimal.
Heuristic [2]	Faster convergence. Leverages problem-specific knowledge.	Limited diversity. Potential bias towards specific regions.
Knowledge-based [3]	High-quality initial solutions. Faster convergence.	Dependence on data quality. Potential bias from outdated knowledge.
Deterministic [4]	Full coverage of the search space. Repeatability.	Lack of diversity. Challenges in high-dimensional problems.
Oversampling [5]	Enhanced diversity. Improved initial solution quality.	Increased computational effort.
Hybrid [6]	Adaptability to diverse problem characteristics.	Increased complexity. Requires careful parameter tuning.

9.1.2 *Selecting the Appropriate Initialization*

The choice of the best initialization technique depends on several factors, including:

- **Problem Characteristics**: The nature of the optimization problem, such as the complexity of the search space, the presence of constraints, and the number of decision variables, plays a significant role in determining the most suitable initialization method. For example, problems with well-structured search spaces may benefit from deterministic initialization, while those with high dimensionality might require oversampling or hybrid approaches to ensure sufficient diversity.

- **Algorithm Characteristics**: Different metaheuristic algorithms have varying levels of exploration and exploitation capabilities. Matching the initialization technique with the algorithm's search behavior is crucial. For instance, algorithms with strong exploitation capabilities might benefit from heuristic or knowledge-based initialization to quickly focus on promising regions, while those with strong exploration capabilities might work well with random or oversampling initialization to cover a wider search space.
- **Computational Resources**: Some initialization techniques, such as oversampling, require more computational resources than others. When dealing with computationally intensive problems or limited resources, it might be necessary to opt for simpler methods like random or heuristic initialization.
- **Desired Balance:** The desired balance between exploration and exploitation also influences the choice of initialization. If extensive exploration is crucial, techniques like random or oversampling are preferable. If faster convergence is a priority, heuristic or knowledge-based initialization might be more appropriate.

In conclusion, the initialization phase is an important step in designing metaheuristic algorithms. It affects how well the algorithm searches for solutions and finds high-quality results. Choosing the right initialization method plays a big role in the algorithm's overall performance, influencing how quickly and accurately those solutions will be found. Different methods, such as random, heuristic, or hybrid approaches, have strengths and weaknesses for optimization problems. Proper initialization helps create a strong starting point for the search process. It reduces the chances of quickly settling on a poor solution and improves the quality of the final results. As problems become more complex, practitioners should carefully examine and adjust their initialization choices as needed. Regularly reviewing this process can help make optimization more efficient and lead to better solutions for challenging real-world issues.

9.2 Diversity Metrics

Diversity metrics are essential in evolutionary algorithms, significantly influencing the search dynamics and performance of the optimization process. They measure population variability, affecting the algorithm's exploration capabilities, convergence behavior, and ability to avoid local optima. By quantifying diversity, these metrics help develop adaptation strategies, ensuring a balance between exploration and exploitation. Despite their benefits, there are challenges associated with their application. This analysis explores the advantages and limitations of employing diversity metrics in evolutionary algorithms, focusing on their effects on performance, computational cost, and the quality of solutions.

Diversity Importance

Evolutionary algorithms thrive on the delicate balance between exploration and exploitation, a dynamic interplay determining their success. Exploration drives the search for novel solutions, encouraging the algorithm to venture into uncharted regions of the solution space while exploitation refines known solutions, optimizing them for better performance. However, these forces can lead to inefficiency or stagnation without proper management. Diversity measures act as mediators, ensuring that the population remains varied enough to explore new possibilities while converging toward optimal solutions. This balance is not static; it evolves as the algorithm progresses, with exploration prioritized early on and exploitation taking precedence later. Diversity measures adapt to this shifting focus, maintaining the algorithm's effectiveness at every stage [7]. Balancing exploration and exploitation is both a science and an art, requiring intuition and creativity to craft innovative yet efficient algorithms. By mastering this balance, evolutionary algorithms unlock their full potential, producing solutions as adaptable and resilient as the natural processes that inspired them.

Types of Diversity

Previously mentioned in Chapters 6 and 7, several strategies and mechanisms exist to assess the diversity among the population of an evolutionary algorithm. These strategies were separated into three main categories for general understanding: Distance-based, Entropy-based, and Variance-based methods. Nonetheless, there exist more categories that are interesting to define. The following list considers different diversities mechanisms:

- **Genotypic Diversity Metrics:** These metrics are formulated to assess how varied the population is based on the genetic representation of the population individuals. The genotype refers to the encoded solution, typically represented as binary strings, real-valued vectors, or permutations, depending on the problem and algorithm. In this category, common metrics are the Hamming distance representing binary solutions and the Euclidean distance for real-valued individuals. Genotypic diversity focuses on the structural differences between individuals, ensuring they remain genetically varied. However, it can be computationally expensive for large populations or high-dimensional problems.
- **Phenotypic Diversity Metrics:** The phenotype emphasizes individuals' observable characteristics or behaviors, such as their fitness values or solution outputs. It reflects these solutions' real significance or impact on the evolutionary process, usually concerning their objective values, behaviors, or functional outputs. Although this metric provides a real representation of the population, it is problem-dependent and must be defined with caution.
- **Niching and Crowding Metrics:** Traditionally, techniques such as niching and crowding are employed in evolutionary algorithms to maintain diversity and encourage exploration of various areas within the search space. Instead

of relying on traditional measures like distances or variances to compute diversity, these methods focus on managing the spatial interactions between individuals during the search process. This approach helps to preserve diversity by regulating how individuals coexist and interact, ultimately enhancing the effectiveness of the search. Niching techniques help develop subpopulations in multimodal environments, while crowding methods ensure that new individuals replace similar ones, preserving diversity and preventing the loss of unique population regions.

- **Hybrid Diversity Metrics:** Hybridization is a common strategy used in evolutionary algorithms, normally involving the integration of different operators or strategies from different algorithms. This conceptualization can be applied in diversity metrics, with the aim to provide a more comprehensive assessment of population diversity by combining genotypic and phenotypic perspectives. These metrics not only assess the genetic differences between individuals but also consider the significant behavioral and performance variations that arise from these genetic differences. However, it requires careful design to combine metrics meaningfully.

9.2.1 *Pros and Cons of Diversity*

Although diversity metrics serve as an important strategy to enhance the evolutionary algorithmic process, they may present limitations. From scalability issues to high computational costs, they can affect the performance of evolutionary algorithms [7], [8]. Table 9.2 depicts the advantages and disadvantages of each defined diversity in this book.

9.2.2 *Selecting the Appropriate Diversity*

When designing a new algorithm or searching for the best configuration optimization framework to solve a problem, it is crucial to carefully select the diversity mechanism to promote a healthy population and a proper balance between exploration and exploitation. The choice of the most appropriate diversity metric depends on several factors, including:

- **Problem Characteristics**: The nature of the optimization problem significantly influences the choice of diversity metric. For example, binary or discrete problems may benefit from genotypic measures like Hamming distance. In contrast, continuous and high-dimensional problems might be better served by metrics such as Euclidean distance or covariance matrices. In multimodal problems, niching-based metrics can help preserve multiple optimal regions, whereas problems focused on behavioral outcomes may require phenotypic diversity measures.
- **Algorithm Characteristics**: Different evolutionary algorithms have unique dynamics and requirements for diversity management. Algorithms with strong

Table 9.2: Advantages and disadvantages of commonly used diversity metrics in evolutionary algorithms.

Diversity	Advantages	Disadvantages
Euclidean Distance [9]	Suitable for real-valued representations. Simple to compute.	Less meaningful in high-dimensional spaces. Sensitive to scaling.
Hamming Distance [10]	Effective for binary representations. Low computational cost.	Limited to discrete representations. Ignores semantic meaning of changes.
Average Population Diversity [11]	Easy to interpret. Provides a general overview of diversity.	May overlook localized diversity patterns. Loses granularity.
Shannon Entropy [12]	Captures positional variability. Good for discrete variables.	Requires discretization for continuous spaces. Ignores structural relationships.
Covariance Matrix [13]	Captures correlations between variables. Useful in continuous domains.	Computationally expensive. Harder to interpret in large dimensions.
Dimension-wise Diversity [14]	Highlights diversity per variable. Helps detect converging dimensions.	Ignores interactions between variables. It may not reflect overall diversity.
Genotypic Diversity [15]	Representation specific. Easy to compute for standard encodings.	May not reflect functional or behavioral differences.
Phenotypic Diversity [16]	Captures meaningful behavioral differences. Representation independent.	Problem-specific. It can be computationally expensive.
Niching Methods [17]	Preserves subpopulations and multiple optima. Supports multimodal problems.	Requires parameter tuning. Higher computational cost.
Crowding Distance [18] [19]	Promotes uniform spread in multi-objective optimization. Simple to implement.	Limited to nearby individuals. Less effective in very large populations.
Hybrid Diversity Metrics [20]	Combines structural and behavioral insights. Adaptive and flexible.	Increased complexity. Requires careful balancing of components.

exploratory behavior may require only simple diversity tracking (such as average population diversity). In contrast, algorithms prone to premature convergence may benefit from more advanced metrics like Shannon entropy or hybrid diversity measures that monitor genetic and functional diversity.

- **Computational Resources**: Some diversity metrics involve higher computational overhead, like covariance matrix calculations or hybrid approaches. For large populations or real-time applications, lightweight metrics such as Hamming distance or crowding distance are often more practical. Based on available resources, the trade-off between metric accuracy and computational cost should be carefully considered.
- **Desired Diversity Control:** The intended role of diversity within the search process also affects the choice of metric. If the goal is to maintain structural variety, genotypic metrics are appropriate. If behavioral variety or solution performance differences are prioritized, phenotypic or niching-based metrics are preferable. Hybrid diversity metrics provide comprehensive monitoring across both dimensions for a balanced approach.

Diversity metrics are important in evolutionary algorithms, helping to balance exploration and exploitation, which affects how well the algorithm performs and how quickly it finds solutions. Choosing the right diversity metric depends on the problem, the algorithm, and the available resources. Different metrics, like genotypic, phenotypic, hybrid, niching, and crowding, each have their own benefits based on the optimization goals. Diversity is key to a successful search process. If ignored, algorithms can get stuck and find poor solutions too soon. It is essential to choose a metric that fits the challenges of the problem to keep the search diverse and productive. Practitioners should start with simple diversity measures to check the health of the population. As problems become more complex, they can use more advanced metrics. Regularly checking diversity and being willing to adjust strategies can greatly improve the success of evolutionary algorithms in different types of environments.

References

[1] W. D. Kelton, "Random initialization methods in simulation," *IIE Transactions*, vol. 21, no. 4, pp. 355–367, 1989.

[2] S. A. Alsaidy, A. D. Abbood, and M. A. Sahib, "Heuristic initialization of pso task scheduling algorithm in cloud computing," *Journal of King Saud University-Computer and Information Sciences*, vol. 34, no. 6, pp. 2370–2382, 2022.

[3] C. Li, X. Chu, Y. Chen, and L. Xing, "A knowledge-based technique for initializing a genetic algorithm," *Journal of Intelligent & Fuzzy Systems*, vol. 31, no. 2, pp. 1145–1152, 2016.

[4] S. Elsayed, R. Sarker, and C. A. C. Coello, "Sequence-based deterministic initialization for evolutionary algorithms," *IEEE Transactions on Cybernetics*, vol. 47, no. 9, pp. 2911–2923, 2016.

[5] S. Sharma, A. Gosain, and S. Jain, "A review of the oversampling techniques in class imbalance problem," in *International Conference on Innovative Computing and Communications: Proceedings of ICICC 2021, Volume 1*. Springer, 2022, pp. 459–472.

[6] G. S. Martínez, T. Karhela, H. Niemistö, A. Rossi, C. Pang, and V. Vyatkin, "A hybrid approach for the initialization of tracking simulation systems," in *2015 IEEE 20th Conference on Emerging Technologies & Factory Automation (ETFA)*. IEEE, 2015, pp. 1–8.

[7] D. Sudholt, "The benefits of population diversity in evolutionary algorithms: a survey of rigorous runtime analyses," *Theory of Evolutionary Computation: Recent Developments in Discrete Optimization*, pp. 359–404, 2019.

[8] M. Pluhacek, A. Viktorin, T. Kadavy, and A. Kazikova, "On the common population diversity measures in metaheuristics and their limitations," in *2021*

IEEE Symposium Series on Computational Intelligence (SSCI). IEEE, 2021, pp. 1–7.

[9] R. Fabbri, L. D. F. Costa, J. C. Torelli, and O. M. Bruno, "2d euclidean distance transform algorithms: A comparative survey," *ACM Computing Surveys (CSUR)*, vol. 40, no. 1, pp. 1–44, 2008.

[10] S. J. Louis and G. J. Rawlins, "Syntactic analysis of convergence in genetic algorithms," in *Foundations of Genetic Algorithms*. Elsevier, 1993, vol. 2, pp. 141–151.

[11] R. Poláková, J. Tvrdík, and P. Bujok, "Differential evolution with adaptive mechanism of population size according to current population diversity," *Swarm and Evolutionary Computation*, vol. 50, p. 100519, 2019.

[12] C. E. Shannon, "A mathematical theory of communication," *The Bell System Technical Journal*, vol. 27, no. 4, pp. 623–656, 1948.

[13] S. Ghosh, S. Das, S. Roy, S. M. Islam, and P. N. Suganthan, "A differential covariance matrix adaptation evolutionary algorithm for real parameter optimization," *Information Sciences*, vol. 182, no. 1, pp. 199–219, 2012.

[14] S. Cheng, Y. Shi, Q. Qin, Q. Zhang, and R. Bai, "Population diversity maintenance in brain storm optimization algorithm," *Journal of Artificial Intelligence and Soft Computing Research*, vol. 4, no. 2, pp. 83–97, 2014.

[15] G. Corriveau, R. Guilbault, A. Tahan, and R. Sabourin, "Review and study of genotypic diversity measures for real-coded representations," *IEEE Transactions on Evolutionary Computation*, vol. 16, no. 5, pp. 695–710, 2012.

[16] ——, "Review of phenotypic diversity formulations for diagnostic tool," *Applied Soft Computing*, vol. 13, no. 1, pp. 9–26, 2013.

[17] E. Yu and P. N. Suganthan, "Ensemble of niching algorithms," *Information Sciences*, vol. 180, no. 15, pp. 2815–2833, 2010.

[18] B. Luo, J. Zheng, J. Xie, and J. Wu, "Dynamic crowding distance? a new diversity maintenance strategy for moeas," in *2008 Fourth International Conference on Natural Computation*, vol. 1. IEEE, 2008, pp. 580–585.

[19] P. Samui, Ed., *Handbook of Research on Advanced Computational Techniques for Simulation-Based Engineering*. IGI Global, 2016.

[20] A. Petrovan, O. Matei, P. C. Pop, and C. Sabo, "Diversity population metrics in diploid and haploid genetic algorithm variants," in *International Conference on Hybrid Artificial Intelligence Systems*. Springer, 2024, pp. 324–338.

Appendix A

Test Functions

In this appendix, the MATLAB® codes for five common test functions are provided: Sphere, Rothyp, Ackley, Multimodal, and Step. These functions can be implemented to test the proposed codes of this book.

A.1 Sphere

```
%   SPHERE function:
%
%   Number or variables(dimensions): n.
%   Search domain: -5.12<=xi<=5.12, i = 1,2,...,n.
%   Global minima: Sphere.Fit(x*) = 0
%   Global solution: x* = (0,...,0).
%   Get bounds: Sphere.Bounds(n);

classdef Sphere
    methods(Static)
        function [fit] = Fit(x)
            if iscolumn(x)
                x = x'; % Ensure x being a row vector
            end
            fit = sum(x.^2);
        end

        function [low, up] = Bounds(dims)
            low = -5.12 * ones(1, dims);
            up = 5.12 * ones(1, dims);
        end
    end
end

% Example of calling Sphere.Fit(x) with a function handle
% SphereFunctionHandle = @(x) Sphere.Fit(x);
```

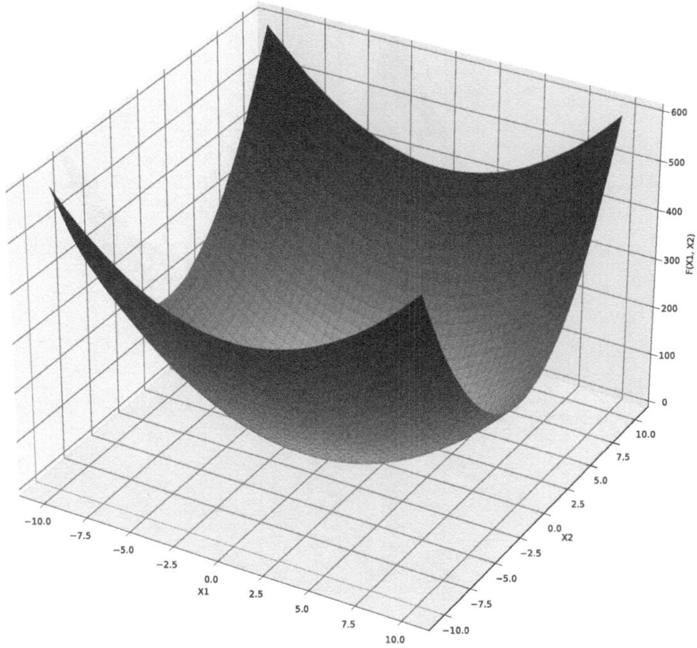

Figure A.1: Sphere function representation.

A.2 Rothyp

```
%   ROTATED HYPER-ELLIPSOID function:
%
%   Number of variables(dimensions): n.
%   Search domain: -65.536<=xi<=65.536, i = 1,2,...,n.
%   Global minima: Rothyp.Fit(x*) = 0
%   Global solution: x* = (0,...0).
%   Get bounds: Rothyp.Bounds(n);

classdef Rothyp
    methods(Static)
        function [fit] = Fit(x)
            if iscolumn(x)
                x = x'; % Ensure x being a row vector
            end
            fit = sum(cumsum(x.^2)); % Vectorized implementation
        end

        function [low, up] = Bounds(dims)
            low = -65.536 * ones(1, dims);
            up = 65.536 * ones(1, dims);
        end
    end
end

% Example of calling Rothyp.Fit(x) with a function handle
% RothypFunctionHandle = @(x) Rothyp.Fit(x);
```

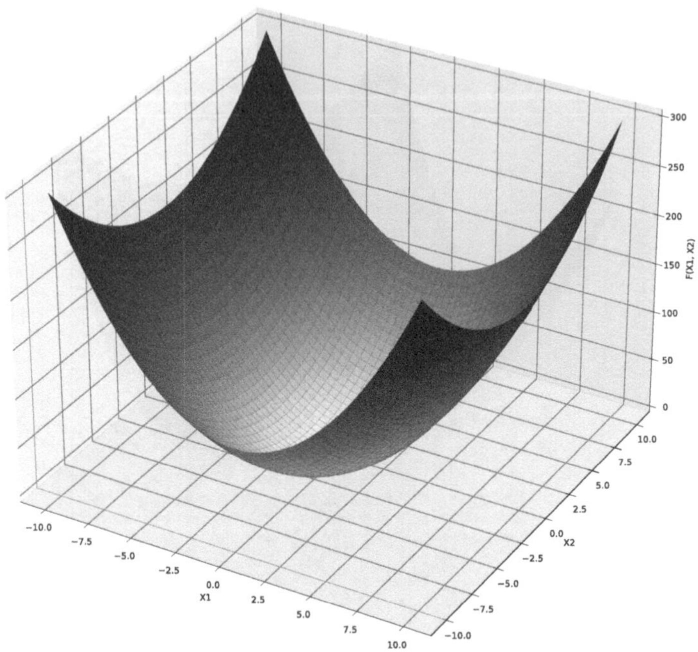

Figure A.2: Rothyp function representation.

A.3 Ackley

```
%   ACKLEY function:
%
%   Number or variables(dimensions): n.
%   Search domain: -30<=xi <=30, i = 1,2,...,n.
%   Global minima: Ackley.Fit(x*) = 0
%   Global solution: x* = (0,...0).
%   Get bounds: Ackley.Bounds(n);

classdef Ackley
    methods(Static)
        function [fit] = Fit(x)
            if iscolumn(x)
                x = x'; % Ensure x being a row vector
            end
            d = numel(x); % Dimensions number
            fit = -20 * exp(-0.2 * sqrt(sum(x.^2) / d)) ...
                - exp(sum(cos(2 * pi * x)) / d) + exp(1) + 20;
        end

        function [low, up] = Bounds(dims)
            low = -30 * ones(1, dims);
            up = 30 * ones(1, dims);
        end
    end
end

% Example of calling Ackley.Fit(x) with a function handle
% AckleyFunctionHandle = @(x) Ackley.Fit(x);
```

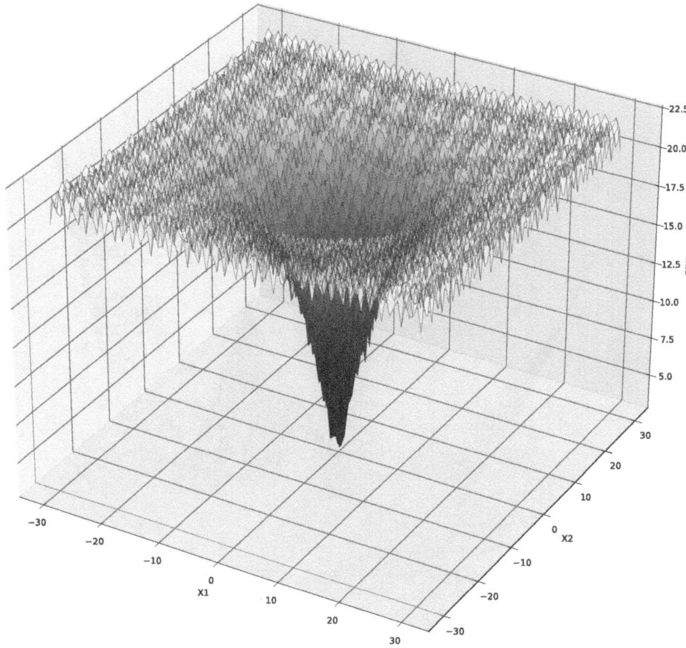

Figure A.3: Ackley function representation.

A.4 Multimodal

```
%    MULTIMODAL function:
%
%    Number or variables(dimensions): n.
%    Search domain: -10<=xi <=10, i = 1,2,...,n.
%    Global minima: MultiModal.Fit(x*) = 0
%    Global solution: x* = (0,...,0).
%    Get bounds: MultiModal.Bounds(n);

classdef MultiModal
    methods(Static)
        function [fit] = Fit(x)
            if iscolumn(x)
                x = x'; % Ensure x being a row vector
            end
            sum_abs = sum(abs(x));
            prod_abs = prod(abs(x));
            fit = sum_abs * prod_abs;
        end

        function [low, up] = Bounds(dims)
            low = -10 * ones(1, dims);
            up = 10 * ones(1, dims);
        end
    end
end

% Example of calling MultiModal.Fit(x) with a function handle
% MultiModalFunctionHandle = @(x) MultiModal.Fit(x);
```

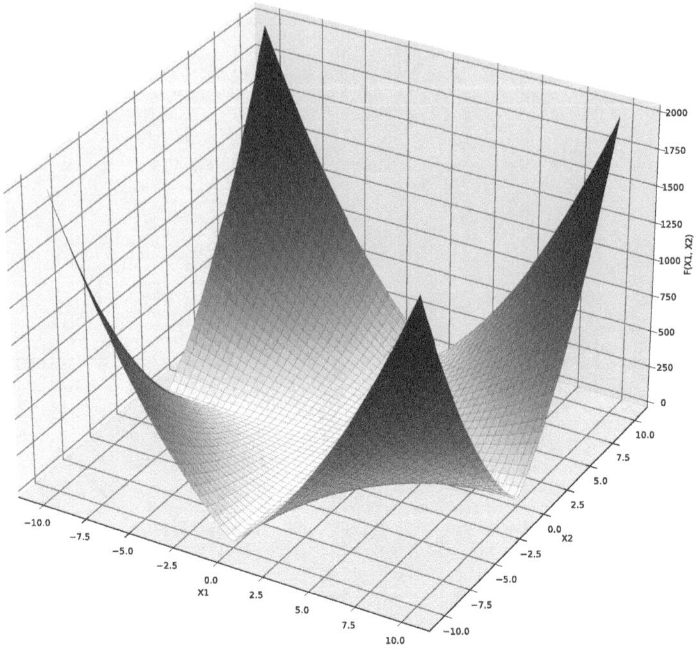

Figure A.4: Multimodal function representation.

A.5 Step

```
%   STEP function:
%
%   Number of variables(dimensions): n.
%   Search domain: -100<=xi<=100, i = 1,2,...,n.
%   Global minima: Step.Fit(x*) = 0
%   Global solution: x* = (0,...,0).
%   Get bounds: Step.Bounds(n);

classdef Step
    methods(Static)
        function [fit] = Fit(x)
            if iscolumn(x)
                x = x'; % Ensure x being a row vector
            end
            fit = sum(floor(x).^2); % Vectorized evaluation
        end

        function [low, up] = Bounds(dims)
            low = -100 * ones(1, dims);
            up = 100 * ones(1, dims);
        end
    end
end

% Example of calling Step.Fit(x) with a function handle
% StepFunctionHandle = @(x) Step.Fit(x);
```

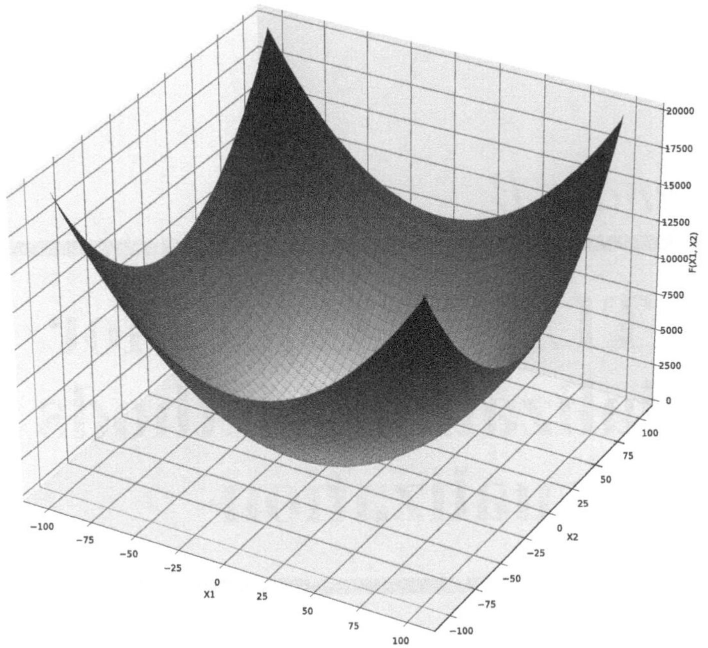

Figure A.5: Step function representation.

A.6 Example Usage of the Ackley Function on the Differential Evolution Algorithm

```
dims = 10; % Number of dimensions
[low_bound, up_bound] = Ackley.Bounds(dims);
AckleyFunctionHandle = @(x) Ackley.Fit(x);
NP = 50;     % Population size
G = 100;     % Number of iterations
F = 0.8;     % Differential weight parameter
CR = 0.9;    % Crossover rate parameter

[best_solution, best_fitness] = DifferentialEvolution(AckleyFunctionHandle, NP,
    G, F, CR, low_bound, up_bound);

disp(['Best found fitness: ', num2str(best_fitness)]);
disp(['best found solution: ', num2str(best_solution)]);
```

Appendix B

MATLAB® Codes for Initialization Methods and 2D Visualization

In this appendix, the MATLAB codes for five common initialization methods are provided: Random initialization, Heuristic initialization, Knowledge initialization, Deterministic initialization, Oversampling initialization, and Hybrid initialization. These methods can be implemented to test and compare different optimization algorithms. Additionally, visualization routines are included to represent the generated solutions in a 2D space.

B.1 Random Initialization

```
clc; clear; close all;

% Parameters for random initialization
N = 100;  % Number of points
dim = 2; % Dimension of the space (2D)
xmin = [0, 0]; % Lower bounds
xmax = [10, 10]; % Upper bounds

% Generate random points uniformly distributed in the search space
points = xmin + (xmax - xmin) .* rand(N, dim);

% Visualization
figure;
scatter(points(:,1), points(:,2), 50, 'filled', 'MarkerFaceColor', '#14acdd');
xlim([xmin(1), xmax(1)]);
ylim([xmin(2), xmax(2)]);
grid on;
xlabel('Dimension 1');
```

```
ylabel('Dimension 2');
%title('Random initialization in 2D');
axis equal;
```

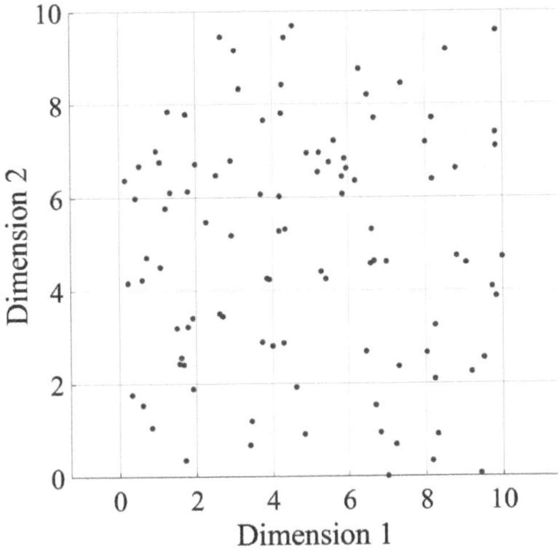

Figure B.1: Random initialization in 2D.

B.2 Heuristic Initialization

```
clc; clear; close all;

% Parameters for heuristic initialization
N = 100; % Number of points
dim = 2; % Dimension of the space (2D)
xmin = [0, 0]; % Lower bounds
xmax = [10, 10]; % Upper bounds

% Generate points using the Halton sequence
points = haltonset(dim, 'Skip', 1e3, 'Leap', 10);
points = net(points, N); % Get the first N points

% Scale points to the desired domain
points = xmin + (xmax - xmin) .* points;

% Visualization
figure;
scatter(points(:,1), points(:,2), 50, 'filled', 'MarkerFaceColor', '#14acdd');
xlim([xmin(1), xmax(1)]);
ylim([xmin(2), xmax(2)]);
grid on;
xlabel('Dimension 1');
ylabel('Dimension 2');
%title('Heuristic initialization using Halton sequence');
axis equal;
```

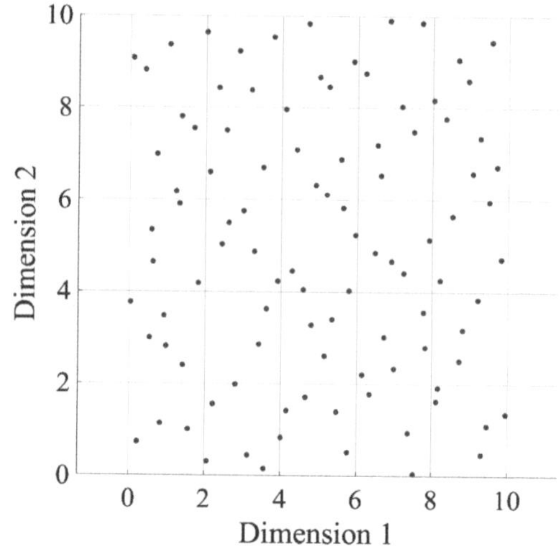

Figure B.2: Heuristic initialization using Halton sequence in 2D.

B.3 Knowledge-based Initialization

```
clc; clear; close all;

% Parameters
N = 100; % Number of points
dim = 2; % Dimension of the space (2D)

% Define knowledge-based centroids (predefined good regions)
centroids = [3 3; 7 7; 5 2]; % Three known good areas
num_clusters = size(centroids, 1);

% Standard deviation for Gaussian spread around centroids
sigma = [1.0, 1.0];

% Generate points based on Gaussian distribution around each centroid
points = [];
for i = 1:num_clusters
    num_points = round(N / num_clusters); % Evenly distribute points
    new_points = centroids(i,:) + sigma .* randn(num_points, dim);
    points = [points; new_points]; % Append points
end

% Visualization
figure;
scatter(points(:,1), points(:,2), 50, 'filled', 'MarkerFaceColor', '#14acdd');
hold on;
scatter(centroids(:,1), centroids(:,2), 100, 'd', 'filled', 'MarkerFaceColor', '
    k'); % Mark centroids
xlim([0, 10]);
ylim([0, 10]);
grid on;
xlabel('Dimension 1');
ylabel('Dimension 2');
%title('Knowledge-Based Initialization in 2D');
legend({'Sampled Points', 'Centroids (Domain Knowledge)'}, 'Location', 'best');
axis equal;
```

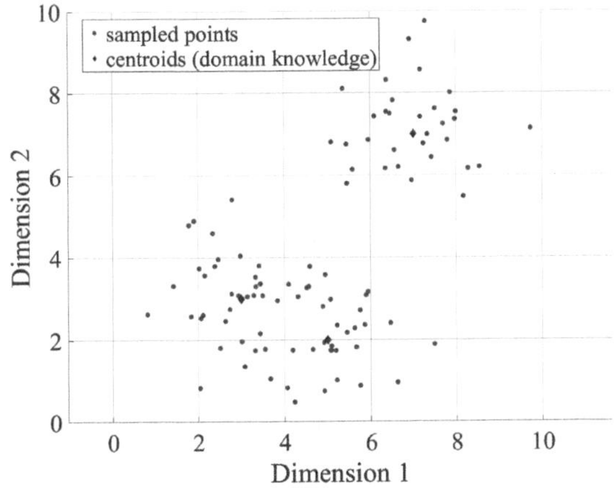

Figure B.3: Knowledge-based initialization in 2D.

B.4 Deterministic Initialization

```
clc; clear; close all;

% Parameters for deterministic initialization
grid_size = 10; % Number of points per axis (grid resolution)
xmin = [0, 0]; % Lower bounds
xmax = [10, 10]; % Upper bounds

% Generate a uniform grid of points
[X, Y] = meshgrid(linspace(xmin(1), xmax(1), grid_size), ...
                  linspace(xmin(2), xmax(2), grid_size));

% Reshape grid into a list of points
points = [X(:), Y(:)];

% Visualization
figure;
scatter(points(:,1), points(:,2), 50, 'filled', 'MarkerFaceColor', '#14acdd');
xlim([xmin(1), xmax(1)]);
ylim([xmin(2), xmax(2)]);
grid on;
xlabel('Dimension 1');
ylabel('Dimension 2');
%title('Deterministic initialization (grid-based) in 2D');
axis equal;
```

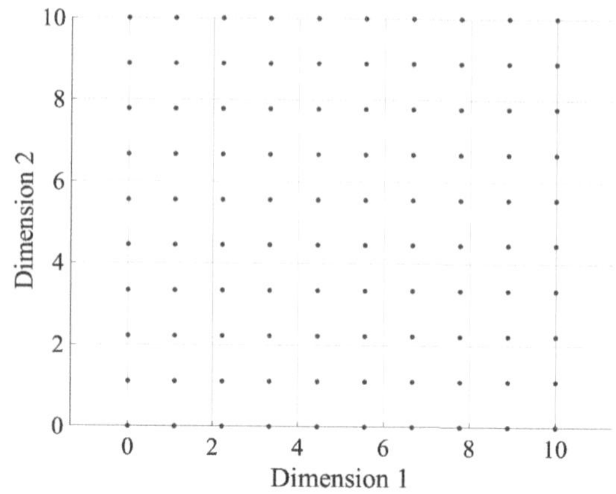

Figure B.4: Deterministic initialization in 2D.

B.5 Oversampling Initialization

```
clc; clear; close all;

% Parameters
N = 100;   % Final number of selected points
oversampling_factor = 3; % How many extra points to generate
N_extra = N * oversampling_factor; % Total initial points
dim = 2;  % Dimension (2D)
xmin = [0, 0]; % Lower bounds
xmax = [10, 10]; % Upper bounds

% Step 1: Generate extra random points
extra_points = xmin + (xmax - xmin) .* rand(N_extra, dim);

% Step 2: Select diverse points using Max-Min distance
selected_points = extra_points(1, :); % Start with first point
for i = 2:N
    % Compute distances from selected points
    distances = pdist2(extra_points, selected_points);
    min_dist = min(distances, [], 2); % Minimum distance to selected points
    [~, idx] = max(min_dist); % Pick the farthest point
    selected_points = [selected_points; extra_points(idx, :)];
end

% Visualization
figure;
scatter(extra_points(:,1), extra_points(:,2), 30, 'MarkerEdgeColor', [0.7 0.7
    0.7]); % Extra points (gray)
hold on;
scatter(selected_points(:,1), selected_points(:,2), 50, 'filled', '
    MarkerFaceColor', '#14acdd'); % Final selected points
xlim([xmin(1), xmax(1)]);
ylim([xmin(2), xmax(2)]);
grid on;
xlabel('Dimension 1');
ylabel('Dimension 2');
%title('Oversampling Initialization (Max-Min Selection)');
legend({'Oversampled solutions', 'Selected subset solutions'}, 'Location', 'best
    ');
axis equal;
```

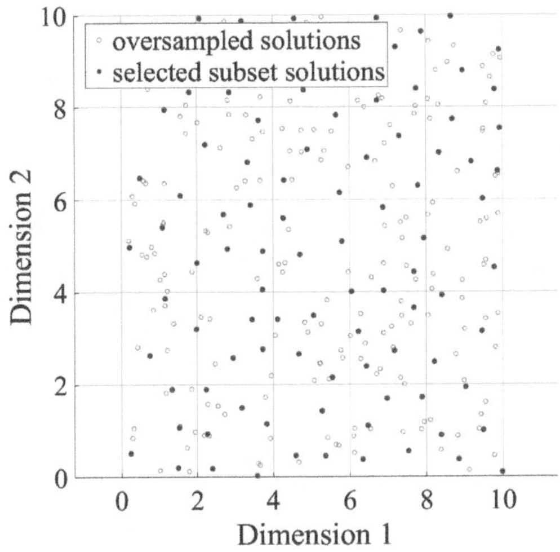

Figure B.5: Oversampling initialization in 2D.

B.6 Hybrid Initialization

```
clc; clear; close all;

% Parameters
N = 100;  % Total number of points
dim = 2; % Dimension (2D)
xmin = [0, 0]; % Lower bounds
xmax = [10, 10]; % Upper bounds

% Distribution of points across different methods
N_random = round(0.4 * N);   % 40% Random Initialization
N_grid = round(0.3 * N);     % 30% Grid-Based Initialization
N_knowledge = N - N_random - N_grid; % Remaining Knowledge-Based

%% Step 1: Random initialization
random_points = xmin + (xmax - xmin) .* rand(N_random, dim);

%% Step 2: Grid-based deterministic initialization
grid_size = ceil(sqrt(N_grid)); % Determine grid resolution
[X, Y] = meshgrid(linspace(xmin(1), xmax(1), grid_size), ...
                  linspace(xmin(2), xmax(2), grid_size));
grid_points = [X(:), Y(:)];
grid_points = grid_points(1:N_grid, :); % Select first N_grid points

%% Step 3: Knowledge-based initialization (clustered gaussian)
centroids = [3 3; 7 7; 5 2]; % Predefined regions
num_clusters = size(centroids, 1);
sigma = [1.0, 1.0]; % Spread of gaussian distribution
knowledge_points = [];

for i = 1:num_clusters
    num_points = round(N_knowledge / num_clusters);
    new_points = centroids(i,:) + sigma .* randn(num_points, dim);
    knowledge_points = [knowledge_points; new_points];
end

%% Combine all initialization methods
hybrid_points = [random_points; grid_points; knowledge_points];
```

```
% Visualization
figure;
hold on;
scatter(random_points(:,1), random_points(:,2), 50, 'filled', 'MarkerFaceColor',
    'r'); % Random (Red)
scatter(grid_points(:,1), grid_points(:,2), 50, 'filled', 'MarkerFaceColor', '
    #14acdd'); % Grid-based (Blue)
scatter(knowledge_points(:,1), knowledge_points(:,2), 50, 'filled', '
    MarkerFaceColor', 'g'); % Knowledge-based (Green)
scatter(centroids(:,1), centroids(:,2), 100, 'd', 'filled', 'MarkerFaceColor', '
    k'); % Knowledge centroids (Black)

xlim([xmin(1), xmax(1)]);
ylim([xmin(2), xmax(2)]);
grid on;
xlabel('Dimension 1');
ylabel('Dimension 2');
%title('Hybrid initialization (Random + Grid + Knowledge-based)');
legend({'Random', 'Grid-based', 'Knowledge-based', 'Knowledge centroids'}, '
    Location', 'best');
axis equal;
hold off;
```

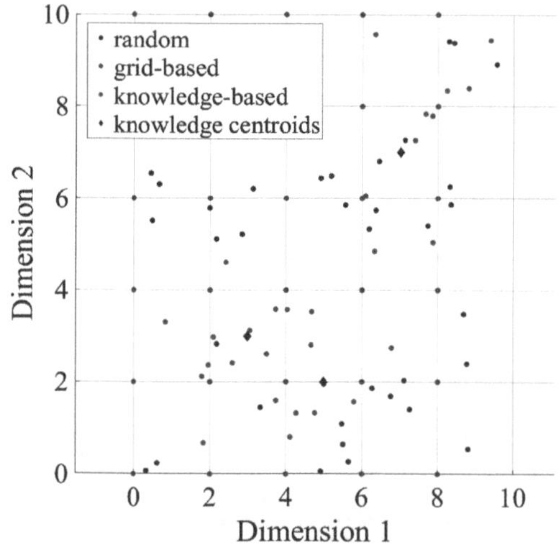

Figure B.6: Hybrid initialization in 2D.

Appendix C

Solutions

C.1 Solutions to Problems of Chapter 4

The solutions are presented here.

Subsection 4.3.1 Practice exercises for Random initialization

(a) Solution to Exercise 1: Visualization of the initial population.

```
% Example of code to solve the proposed Exercise 1
clc; clear; close all;

% Define the problem
n = 100; % Number of points (solutions)
d = 2; % Number of dimensions (variables)
Xmin = [0, 0]; % Lower bounds for each variable
Xmax = [10, 20]; % Upper bounds for each variable

% Random Initialization
Pop = zeros(n, d); % Preallocate population matrix
for i = 1:n
    for j = 1:d
        Pop(i, j) = Xmin(j) + (Xmax(j) - Xmin(j)) * rand();
    end
end

% Plot the randomly initialized solutions
figure;
scatter(Pop(:,1), Pop(:,2), 'filled');
xlabel('Variable 1');
ylabel('Variable 2');
title('Random Initialization in 2D');
grid on;
```

(b) Solution to Exercise 2: Statistical analysis of the distribution of solutions.

```
% Example of code to solve the proposed Exercise 2
clc; clear; close all;

% Parameters definition
n = 500; % Number of solutions
Xmin = [0, 0, 0]; % Lower bounds for each dimension
Xmax = [10, 20, 30]; % Upper bounds for each dimension
d = length(Xmin); % Number of dimensions

% Generate the random population
Pop = rand(n, d) .* (Xmax - Xmin) + Xmin;

% Calculate mean and standard deviation for each dimension
mean_values = mean(Pop);
std_values = std(Pop);

% Display statistical results
fprintf('Mean values per dimension: \n');
disp(mean_values);
fprintf('Standard deviation per dimension: \n');
disp(std_values);

% Boxplot visualization
figure;
boxplot(Pop, 'Labels', {'Dim 1', 'Dim 2', 'Dim 3'});
title('Boxplot of solution distribution');
xlabel('Dimensions');
ylabel('Values');
grid on;

% Histogram comparison with uniform theoretical distribution
figure;
for i = 1:d
    subplot(1, d, i);
    histogram(Pop(:, i), 'Normalization', 'pdf');
    hold on;
    % Theoretical uniform distribution
    yline(1 / (Xmax(i) - Xmin(i)), 'r', 'LineWidth', 1.5);
    title(['Histogram of Dimensions ', num2str(i)]);
    xlabel('Values');
    ylabel('Density');
    grid on;
end
title('Histogram comparison with random uniform distribution');
```

Subsection 4.3.2 Practice exercises for Heuristic initialization

(a) Solution to Exercise 1: Implementation of heuristic initialization in a routing problem.

```
% Example of code to solve the proposed Exercise 1
clc; clear; close all;

% Number of cities
numCities = 50;

% Generate random coordinates in a 2D space
cities = rand(numCities, 2) * 100;

% Choose a random starting city
startCity = randi(numCities);
```

```
% Apply the nearest neighbor heuristic
visited = false(numCities, 1);
visited(startCity) = true;
route = startCity;

currentCity = startCity;

for i = 2:numCities
    % Find the nearest unvisited city
    distances = vecnorm(cities - cities(currentCity, :), 2, 2);
    distances(visited) = inf; % Avoid already visited cities
    [~, nextCity] = min(distances);

    % Update route and mark as visited
    route = [route, nextCity];
    visited(nextCity) = true;
    currentCity = nextCity;
end

% Return to the starting city to close the circuit
route = [route, startCity];

% Visualize the generated route
figure;
plot(cities(route,1), cities(route,2), '-o', 'LineWidth', 1.5, 'MarkerSize', 8);
hold on;
scatter(cities(:,1), cities(:,2), 50, 'filled'); % City points
title('Initial route using the nearest neighbor heuristic');
xlabel('X Coordinate');
ylabel('Y Coordinate');
grid on;
```

(b) Solution to Exercise 2: Evaluation of diversity in heuristic initialization.

```
% Example of code to solve the proposed Exercise 2
clc; clear; close all;

% Number of solutions
n = 100;

% Search space boundaries
Xmin = [0, 0];
Xmax = [50, 100];

% Random initialization
random_solutions = Xmin + rand(n, 2) .* (Xmax - Xmin);

% Heuristic initialization (based on an estimated optimal region)
optimal_region = [25, 50]; % Estimated promising region
heuristic_solutions = optimal_region + randn(n, 2) * 5; % Centered distribution

% Visualize both distributions
figure;
subplot(1,2,1);
scatter(random_solutions(:,1), random_solutions(:,2), 20, 'b', 'filled');
title('Random initialization distribution');
xlabel('X Dimension');
ylabel('Y Dimension');
xlim([Xmin(1), Xmax(1)]);
ylim([Xmin(2), Xmax(2)]);
grid on;

subplot(1,2,2);
scatter(heuristic_solutions(:,1), heuristic_solutions(:,2), 20, 'r', 'filled');
```

```
title('Heuristic initialization distribution');
xlabel('X Dimension');
ylabel('Y Dimension');
xlim([Xmin(1), Xmax(1)]);
ylim([Xmin(2), Xmax(2)]);
grid on;

% Compute the average distance between solutions in each set
dist_random = pdist(random_solutions);
dist_heuristic = pdist(heuristic_solutions);

fprintf('Average distance in random initialization: %.2f\n', mean(dist_random));
fprintf('Average distance in heuristic initialization: %.2f\n', mean(
    dist_heuristic));
```

Subsection 4.3.2 Practice exercises for Knowledge-based initialization

(a) Solution to Exercise 1: Knowledge-based initialization for vehicle routing optimization.

```
%% Exercise 1: Knowledge-Based Initialization for Vehicle Routing Optimization
clc; clear; close all;

% Define 10 customer locations in a 2D space
num_customers = 100;
customers = rand(num_customers, 2) * 100;

% Generate historical routes (for simplicity, assume a predefined order)
historical_routes = [1:num_customers; randperm(num_customers)];

% Plot historical route
figure; hold on;
scatter(customers(:,1), customers(:,2), 'bo', 'filled');
plot(customers(historical_routes(1,:), 1), customers(historical_routes(1,:), 2),
    'r-', 'LineWidth', 2);
title('Knowledge-Based Vehicle Routing Initialization');
xlabel('X Coordinate'); ylabel('Y Coordinate'); grid on;
legend('Customers', 'Historical Route');

% Extra Challenge: Compare with Random Initialization
random_route = randperm(num_customers);
distance_historical = sum(vecnorm(diff(customers(historical_routes(1,:), :))')');
distance_random = sum(vecnorm(diff(customers(random_route, :))'));

fprintf('Total distance (historical route): %.2f\n', distance_historical);
fprintf('Total distance (random route): %.2f\n', distance_random);
```

(b) Solution to Exercise 2: Expert knowledge-based scheduling optimization.

```
%% Exercise 2: Expert Knowledge for Production Scheduling
clc; clear; close all;

% Define machines and tasks
num_machines = 5;
num_tasks = 15;
processing_times = randi([5, 20], num_tasks, 1); % Random processing times

% Expert Heuristic: Sort tasks by shortest processing time first
[sorted_times, sorted_indices] = sort(processing_times);

% Assign tasks to machines in a round-robin fashion
schedule = cell(num_machines, 1);
```

```
for i = 1:num_tasks
    machine_idx = mod(i-1, num_machines) + 1;
    schedule{machine_idx} = [schedule{machine_idx}, sorted_indices(i)];
end

% Visualization of Initial Scheduling
figure; hold on;
for m = 1:num_machines
    y = m * ones(size(schedule{m}));
    x = sorted_times(ismember(sorted_indices, schedule{m}));
    scatter(x, y, 'filled');
end

title('Expert-Based production scheduling initialization');
xlabel('Processing Time'); ylabel('Machine ID'); grid on;
legend(arrayfun(@(x) sprintf('Machine %d', x), 1:num_machines, 'UniformOutput',
    false));

% Extra Challenge: Calculate standard deviation of task completion times
task_completion_times = cellfun(@sum, schedule);
std_dev = std(task_completion_times);
fprintf('Standard deviation of completion times: %.2f\n', std_dev);
```

Subsection 4.3.4 Practice exercises for Deterministic initialization

(a) Solution to Exercise 1: Grid-based deterministic initialization.

```
%% Exercise 1: Grid-Based Deterministic Initialization
% Define the 2D search space bounds
X_min = [-100, -100];
X_max = [100, 100];

% Define the grid size (30x30 grid)
grid_size = 30;

% Generate grid-based population
[x_grid, y_grid] = meshgrid(linspace(X_min(1), X_max(1), grid_size), ...
                            linspace(X_min(2), X_max(2), grid_size));

population_grid = [x_grid(:), y_grid(:)];

% Plot the generated population
figure;
scatter(population_grid(:,1), population_grid(:,2), 'filled');
grid on;
xlabel('X-axis'); ylabel('Y-axis');
title('Grid-Based Initialization');

% Extra Challenge: Compute pairwise distances and compare with random
    initialization
random_population = X_min + (X_max - X_min) .* rand(25,2);
distances_grid = pdist(population_grid);
distances_random = pdist(random_population);

fprintf('Mean pairwise distance (Grid): %.4f\n', mean(distances_grid));
fprintf('Mean pairwise distance (Random): %.4f\n', mean(distances_random));
```

(b) Solution to Exercise 2: Sobol sequence-based deterministic initialization.

```
%% Exercise 2: Sobol sequence-based deterministic initialization
% Define the 3D search space bounds
X_min = [0, 0, 0];
X_max = [1, 1, 1];
```

```
% Number of points to generate
n_points = 100;

% Generate Sobol sequence
p = sobolset(3); % 3D Sobol sequence
sobol_points = net(p, n_points);

% Scale the points to fit within the defined bounds
sobol_population = X_min + (X_max - X_min) .* sobol_points;

% Plot the Sobol sequence-based population
figure;
scatter3(sobol_population(:,1), sobol_population(:,2), sobol_population(:,3), '
    filled');
grid on;
xlabel('X-axis'); ylabel('Y-axis'); zlabel('Z-axis');
title('Sobol sequence-based initialization');

% Extra Challenge: Compute average minimum distance between points
D = squareform(pdist(sobol_population));
D(D==0) = inf; % Exclude self-distances
mean_min_dist_sobol = mean(min(D, [], 2));

% Compare with random initialization
random_population = X_min + (X_max - X_min) .* rand(n_points,3);
D_rand = squareform(pdist(random_population));
D_rand(D_rand==0) = inf;
mean_min_dist_random = mean(min(D_rand, [], 2));

fprintf('Mean minimum distance (Sobol): %.4f\n', mean_min_dist_sobol);
fprintf('Mean minimum distance (Random): %.4f\n', mean_min_dist_random);
```

Subsection 4.3.5 Practice exercises for Oversampling initialization

(a) Solution to Exercise 1: Fitness-based oversampling initialization.

```
%% Exercise 1: Fitness-based oversampling initialization

% Define search space bounds
X_min = [-100, -100];
X_max = [100, 100];

% Parameters
m = 200; % Number of candidate solutions
n = 50; % Number of selected solutions

% Generate random candidate solutions
Candidates = X_min + (X_max - X_min) .* rand(m, 2);

% Define fitness function (minimizing distance to [5,5])
fitness = sum((Candidates - [5, 5]).^2, 2);

% Select the top n solutions based on fitness
[~, idx] = sort(fitness);
Selected = Candidates(idx(1:n), :);

% Visualization
figure;
scatter(Candidates(:,1), Candidates(:,2), 'b', 'filled'); hold on;
scatter(Selected(:,1), Selected(:,2), 'r', 'filled');
legend('Candidate Solutions', 'Selected Solutions');
xlabel('X'); ylabel('Y'); title('Fitness-Based Oversampling Initialization');
grid on;
```

(b) Solution to Exercise 2: Diversity-based oversampling initialization.

```
%% Exercise 2: Diversity-Based Oversampling Initialization

% Define search space bounds
X_min = [-30, -30];
X_max = [30, 30];

% Parameters
m = 500; % Number of candidate solutions
n = 100; % Number of selected solutions


% Generate random candidate solutions
Candidates = X_min + (X_max - X_min) .* rand(m, 2);

% Initialize the selected set
Selected = Candidates(randi(m, 1), :);

for i = 2:n
    max_dist = -inf;
    best_candidate = [];

    for j = 1:m
        candidate = Candidates(j, :);
        if ismember(candidate, Selected, 'rows')
            continue;
        end

        % Compute minimum pairwise distance to selected set
        distances = sqrt(sum((Selected - candidate).^2, 2));
        min_distance = min(distances);

        % Select candidate maximizing the minimum distance
        if min_distance > max_dist
            max_dist = min_distance;
            best_candidate = candidate;
        end
    end

    % Add best candidate to selected set
    Selected = [Selected; best_candidate];
end

% Visualization
figure;
scatter(Candidates(:,1), Candidates(:,2), 'b', 'filled'); hold on;
scatter(Selected(:,1), Selected(:,2), 'g', 'filled');
legend('Candidate Solutions', 'Selected Solutions');
xlabel('X'); ylabel('Y'); title('Diversity-based oversampling initialization');
grid on;
```

Subsection 4.3.6 Practice exercises for Hybrid initialization

(a) Solution to Exercise 1: Hybrid initialization with fitness and random sampling.

```
%% Exercise 1: Hybrid initialization with fitness and random sampling
clc; clear; close all;

% Define search space bounds
X_min = [0, 0];
X_max = [10, 10];
m = 50; % Total population size
```

```
% Generate 70% random solutions
random_size = round(0.7 * m);
random_solutions = X_min + (X_max - X_min) .* rand(random_size, 2);

% Generate 30 additional random solutions and select the best 30%
extra_size = 30;
extra_solutions = X_min + (X_max - X_min) .* rand(extra_size, 2);
fitness_values = (extra_solutions(:,1) - 5).^2 + (extra_solutions(:,2) - 5).^2;
[~, sorted_idx] = sort(fitness_values);
selected_solutions = extra_solutions(sorted_idx(1:round(0.3 * m)), :);

% Merge subsets
final_population = [random_solutions; selected_solutions];

% Visualization
figure;
scatter(random_solutions(:,1), random_solutions(:,2), 'bo'); hold on;
scatter(selected_solutions(:,1), selected_solutions(:,2), 'ro', 'filled');
xlabel('x'); ylabel('y'); title('Hybrid Initialization: Random + Fitness-Based')
    ;
legend({'Random Solutions', 'Fitness-Based Solutions'});
grid on;

% Compare mean fitness with pure random initialization
random_fitness = (random_solutions(:,1) - 5).^2 + (random_solutions(:,2) - 5)
    .^2;
hybrid_fitness = (final_population(:,1) - 5).^2 + (final_population(:,2) - 5)
    .^2;
fprintf('Mean Fitness - Random: %.4f\n', mean(random_fitness));
fprintf('Mean Fitness - Hybrid: %.4f\n', mean(hybrid_fitness));
```

(b) Solution to Exercise 2: Hybrid initialization with random, grid-based, and heuristic sampling.

```
%% Exercise 2: Hybrid Initialization with Random, Grid-based, and Heuristic
    Sampling
clc; clear; close all;

% Define search space bounds
X_min = [0, 0];
X_max = [10, 10];
m = 200; % Total population size

% Generate 40% random solutions
random_size = round(0.4 * m);
random_solutions = X_min + (X_max - X_min) .* rand(random_size, 2);

% Generate 30% grid-based solutions
grid_size = round(sqrt(0.3 * m));
grid_x = linspace(X_min(1), X_max(1), grid_size);
grid_y = linspace(X_min(2), X_max(2), grid_size);
[grid_X, grid_Y] = meshgrid(grid_x, grid_y);
grid_solutions = [grid_X(:), grid_Y(:)];
grid_solutions = grid_solutions(1:round(0.3 * m), :);

% Generate 30% heuristic-based solutions (k-means cluster centers)
kmeans_size = round(0.3 * m);
kmeans_points = X_min + (X_max - X_min) .* rand(2 * kmeans_size, 2);
[idx, cluster_centers] = kmeans(kmeans_points, kmeans_size);

% Merge subsets
final_population = [random_solutions; grid_solutions; cluster_centers];

% Visualization
figure;
```

```
scatter(random_solutions(:,1), random_solutions(:,2), 'bo'); hold on;
scatter(grid_solutions(:,1), grid_solutions(:,2), 'gs', 'filled');
scatter(cluster_centers(:,1), cluster_centers(:,2), 'rd', 'filled');
xlabel('x'); ylabel('y'); title('Hybrid initialization: Random + Grid +
    Heuristic');
legend({'Random', 'Grid', 'Heuristic'});
grid on;

% Compare diversity with pure random initialization
pairwise_distances = pdist(final_population);
avg_min_distance = mean(min(squareform(pairwise_distances) + diag(inf(m,1))));
random_pairwise_distances = pdist(random_solutions);
avg_min_distance_random = mean(min(squareform(random_pairwise_distances) + diag(
    inf(random_size,1))));

fprintf('Average Min Distance - Random: %.4f\n', avg_min_distance_random);
fprintf('Average Min Distance - Hybrid: %.4f\n', avg_min_distance);
```

C.2 Solutions to Problems of Chapter 8

The solutions are presented here.

Subsection 8.3 Practice exercises for Implementation of diversity indicators in DE and PSO

(a) Solution to Exercise 1: Plot the diversity metric across iterations.

```
function [best_individual, best_fitness, diversity_log] = DifferentialEvolution(
    f, NP, G, F, CR, lower_bound, upper_bound)
    % f: Objective function handle
    % NP: Number of individuals in the population
    % G: Maximum number of generations
    % F: Differential weight
    % CR: Crossover rate
    % lower_bound: Lower bound of the search space
    % upper_bound: Upper bound of the search space

    % Initialize population
    dimensions = length(lower_bound);
    population = lower_bound + (upper_bound - lower_bound) .* rand(NP,
     dimensions);
    fitness = arrayfun(f, population);

    % Initialize diversity tracking
    highestDiversity = 0;
    diversity_log = zeros(1, G); % Store diversity values over generations

    % Compute initial diversity
    dimensionalMedians = zeros(1, dimensions);
    for j = 1:dimensions
        dimensionalPositions = population(:, j);
        dimensionalMedians(j) = median(dimensionalPositions);
    end

    averageDistance = zeros(1, dimensions);
    for j = 1:dimensions
        distancetoMedian = sum(abs(population(:, j) - dimensionalMedians(j)));
        averageDistance(j) = distancetoMedian / NP;
    end

    dimensionalDiversity = sum(averageDistance);
    averageDiversity = dimensionalDiversity / dimensions;
```

```
highestDiversity = averageDiversity;

% Main loop
for generation = 1:G
    for i = 1:NP
        % Mutation: Create a mutant vector
        indices = randperm(NP, 3);
        while any(indices == i)
            indices = randperm(NP, 3);
        end
        r1 = indices(1);
        r2 = indices(2);
        r3 = indices(3);
        mutant_vector = population(r1, :) + F * (population(r2, :) -
population(r3, :));

        % Ensure the mutant vector is within the bounds
        mutant_vector = max(mutant_vector, lower_bound);
        mutant_vector = min(mutant_vector, upper_bound);

        % Crossover: Create a trial vector
        trial_vector = population(i, :);
        j_rand = randi(dimensions);
        for j = 1:dimensions
            if rand <= CR || j == j_rand
                trial_vector(j) = mutant_vector(j);
            end
        end

        % Selection: Select the better between the trial vector and the
target vector
        trial_fitness = f(trial_vector);
        if trial_fitness < fitness(i)
            population(i, :) = trial_vector;
            fitness(i) = trial_fitness;
        end
    end

    % Update diversity
    for j = 1:dimensions
        dimensionalMedians(j) = median(population(:, j));
    end

    for j = 1:dimensions
        averageDistance(j) = sum(abs(population(:, j) - dimensionalMedians(j
))) / NP;
    end

    dimensionalDiversity = sum(averageDistance);
    averageDiversity = dimensionalDiversity / dimensions;
    if averageDiversity > highestDiversity
        highestDiversity = averageDiversity;
    end
    percentualDiversity = (averageDiversity * 100) / highestDiversity;
    diversity_log(generation) = percentualDiversity;
end

% Output the best solution found
[best_fitness, idx] = min(fitness);
best_individual = population(idx, :);

% Plot diversity over generations
figure;
plot(1:G, diversity_log, 'b', 'LineWidth', 2);
xlabel('Generation');
ylabel('Percentual Diversity');
title('Diversity Metric Across Generations');
```

```
    grid on;
end
```

(b) Solution to Exercise 2: Experiment with alternative ways of measuring diversity.

```
function [best_individual, best_fitness, diversity_log_std,
    diversity_log_pairwise] = DifferentialEvolution(f, NP, G, F, CR,
    lower_bound, upper_bound)
   % f: Objective function handle
   % NP: Number of individuals in the population
   % G: Maximum number of generations
   % F: Differential weight
   % CR: Crossover rate
   % lower_bound: Lower bound of the search space
   % upper_bound: Upper bound of the search space

   % Initialize population
   dimensions = length(lower_bound);
   population = lower_bound + (upper_bound - lower_bound) .* rand(NP,
    dimensions);
   fitness = arrayfun(f, population);

   % Initialize diversity tracking
   highestDiversity_std = 0;
   highestDiversity_pairwise = 0;
   diversity_log_std = zeros(1, G); % Store std-dev-based diversity
   diversity_log_pairwise = zeros(1, G); % Store pairwise distance diversity

   % Function to calculate standard deviation-based diversity
   function diversity = computeStdDiversity(pop)
       std_deviation = std(pop, 0, 1);
       diversity = sum(std_deviation) / dimensions;
   end

   % Function to calculate pairwise distance-based diversity
   function diversity = computePairwiseDiversity(pop)
       pairwiseDistances = pdist(pop); % Compute pairwise distances
       diversity = mean(pairwiseDistances);
   end

   % Compute initial diversity
   stdDiversity = computeStdDiversity(population);
   pairwiseDiversity = computePairwiseDiversity(population);
   highestDiversity_std = stdDiversity;
   highestDiversity_pairwise = pairwiseDiversity;

   % Main loop
   for generation = 1:G
       for i = 1:NP
           % Mutation: Create a mutant vector
           indices = randperm(NP, 3);
           while any(indices == i)
               indices = randperm(NP, 3);
           end
           r1 = indices(1);
           r2 = indices(2);
           r3 = indices(3);
           mutant_vector = population(r1, :) + F * (population(r2, :) -
    population(r3, :));

           % Ensure the mutant vector is within the bounds
           mutant_vector = max(mutant_vector, lower_bound);
           mutant_vector = min(mutant_vector, upper_bound);

           % Crossover: Create a trial vector
```

```
            trial_vector = population(i, :);
            j_rand = randi(dimensions);
            for j = 1:dimensions
                if rand <= CR || j == j_rand
                    trial_vector(j) = mutant_vector(j);
                end
            end

            % Selection: Select the better between the trial vector and the
 target vector
            trial_fitness = f(trial_vector);
            if trial_fitness < fitness(i)
                population(i, :) = trial_vector;
                fitness(i) = trial_fitness;
            end
        end

        % Compute diversity after the generation update
        stdDiversity = computeStdDiversity(population);
        pairwiseDiversity = computePairwiseDiversity(population);

        if stdDiversity > highestDiversity_std
            highestDiversity_std = stdDiversity;
        end
        if pairwiseDiversity > highestDiversity_pairwise
            highestDiversity_pairwise = pairwiseDiversity;
        end

        percentualDiversity_std = (stdDiversity * 100) / highestDiversity_std;
        percentualDiversity_pairwise = (pairwiseDiversity * 100) /
 highestDiversity_pairwise;
        diversity_log_std(generation) = percentualDiversity_std;
        diversity_log_pairwise(generation) = percentualDiversity_pairwise;
    end

    % Output the best solution found
    [best_fitness, idx] = min(fitness);
    best_individual = population(idx, :);

    % Plot diversity over generations
    figure;
    plot(1:G, diversity_log_std, 'r', 'LineWidth', 2);
    hold on;
    plot(1:G, diversity_log_pairwise, 'g', 'LineWidth', 2);
    xlabel('Generation');
    ylabel('Percentual Diversity');
    title('Comparison of Diversity Metrics Across Generations');
    legend('Std Dev Diversity', 'Pairwise Distance Diversity');
    grid on;
end
```

(c) Solution to Exercise 5: Modify the stopping condition of the DE and the PSO algorithms to include diversity.

```
function [best_individual, best_fitness] = DifferentialEvolution(f, NP, G, F, CR
    , lower_bound, upper_bound)
    % f: Objective function handle
    % NP: Number of individuals in the population
    % G: Maximum number of generations
    % F: Differential weight
    % CR: Crossover rate
    % lower_bound: Lower bound of the search space
    % upper_bound: Upper bound of the search space

    % Initialize population
```

```
dimensions = length(lower_bound);
population = lower_bound + (upper_bound - lower_bound) .* rand(NP,
 dimensions);
fitness = arrayfun(f, population);

% Initialize diversity
highestDiversity = 0;
dimensionalMedians = zeros(1,dimensions);
for j=1:dimensions
    dimensionalPositions = zeros(NP,1);
    for i=1:NP
        dimensionalPositions(i) = population(i, j);
    end
    dimensionalMedians(j)=median(dimensionalPositions);
end

averageDistance = zeros(1,dimensions);
for j=1:dimensions
    distancetoMedian = 0;
    for i=1:NP
        distancetoMedian = distancetoMedian + abs(population(i, j) -
 dimensionalMedians(j));
    end
    averageDistance(j)=distancetoMedian/NP;
end

dimensionalDiversity=sum(averageDistance(:));
averageDiversity = dimensionalDiversity/dimensions;
if averageDiversity > highestDiversity
    highestDiversity = averageDiversity;
end

% Main loop
for generation = 1:G
    for i = 1:NP
        % Mutation: Create a mutant vector
        indices = randperm(NP, 3);
        while any(indices == i)
            indices = randperm(NP, 3);
        end
        r1 = indices(1);
        r2 = indices(2);
        r3 = indices(3);
        mutant_vector = population(r1, :) + F * (population(r2, :) -
 population(r3, :));

        % Ensure the mutant vector is within the bounds
        mutant_vector = max(mutant_vector, lower_bound);
        mutant_vector = min(mutant_vector, upper_bound);

        % Crossover: Create a trial vector
        trial_vector = population(i, :);
        j_rand = randi(dimensions);
        for j = 1:dimensions
            if rand <= CR || j == j_rand
                trial_vector(j) = mutant_vector(j);
            end
        end

        % Selection: Select the better between the trial vector and the
 target vector
        trial_fitness = f(trial_vector);
        if trial_fitness < fitness(i)
            population(i, :) = trial_vector;
            fitness(i) = trial_fitness;
        end
    end
```

```
        % Update diversity
        dimensionalMedians = zeros(1,dimensions);
        for j=1:dimensions
            dimensionalPositions = zeros(NP,1);
            for i=1:NP
                dimensionalPositions(i) = population(i, j);
            end
            dimensionalMedians(j)=median(dimensionalPositions);
        end

        averageDistance = zeros(1,dimensions);
        for j=1:dimensions
            distancetoMedian = 0;
            for i=1:NP
                distancetoMedian = distancetoMedian + abs(population(i, j) -
    dimensionalMedians(j));
            end
            averageDistance(j)=distancetoMedian/NP;
        end

        dimensionalDiversity=sum(averageDistance(:));
        averageDiversity = dimensionalDiversity/dimensions;
        if averageDiversity > highestDiversity
            highestDiversity = averageDiversity;
        end
        percentualDiversity = averageDiversity*100/highestDiversity;

        % Terminate the process if the population diversity is less than 2
    percent
        if percentualDiversity < 2
            break;
        end

    end

    % Output the best solution found
    [best_fitness, idx] = min(fitness);
    best_individual = population(idx, :);
end
```

```
function [global_best_individual, global_best_fitness] =
    ParticleSwarmOptimization(f, NP, T, w, c1, c2, lower_bound, upper_bound)
    % f: Objective function handle
    % NP: Number of particles
    % T: Maximum number of iterations
    % w: Inertia weight
    % c1: Cognitive constant
    % c2: Social constant
    % lower_bound: Lower bound of the search space
    % upper_bound: Upper bound of the search space


    % Initialize particle positions and velocities
    dimensions = length(lower_bound);
    position = lower_bound + (upper_bound - lower_bound) .* rand(NP, dimensions)
    ;
    personal_best_position = position;
    velocity = zeros(NP, dimensions);
    fitness = arrayfun(f, position);

    % Initialize global best
    [global_best_fitness, idx] = min(fitness);
    global_best_individual = personal_best_position(idx, :);

    % Initialize diversity
    highestDiversity = 0;
```

```
dimensionalMedians = zeros(1,dims);
for j=1:dims
    dimensionalPositions = zeros(NP,1);
    for i=1:NP
        dimensionalPositions(i) = position(i, j);
    end
    dimensionalMedians(j)=median(dimensionalPositions);
end

averageDistance = zeros(1,dimensions);
for j=1:dimensions
    distancetoMedian = 0;
    for i=1:NP
        distancetoMedian = distancetoMedian + abs(position(i, j) -
 dimensionalMedians(j));
    end
    averageDistance(j)=distancetoMedian/NP;
end

dimensionalDiversity=sum(averageDistance(:));
averageDiversity = dimensionalDiversity/dimensions;
if averageDiversity > highestDiversity
    highestDiversity = averageDiversity;
end

% Main loop
for t = 1:T
    for i = 1:NP
        % Update velocity
        r1 = rand(1, dimensions);
        r2 = rand(1, dimensions);
        velocity(i, :) = w * velocity(i, :) + ...
                         c1 * r1 .* (personal_best_position(i, :) - position
 (i, :)) + ...
                         c2 * r2 .* (global_best_individual - position(i, :)
 );

        % Update position
        position(i, :) = position(i, :) + velocity(i, :);

        % Ensure the position is within the bounds
        position(i, :) = max(position(i, :), lower_bound);
        position(i, :) = min(position(i, :), upper_bound);

        % Evaluate the new position
        current_fitness = f(position(i, :));

        % Update the particle's best known position
        if current_fitness < fitness(i)
            personal_best_position(i, :) = position(i, :);
            fitness(i) = current_fitness;
        end

        % Update the global best known position
        if current_fitness < global_best_fitness
            global_best_individual = position(i, :);
            global_best_fitness = current_fitness;
        end
    end

    % Update diversity
    highestDiversity = 0;
    dimensionalMedians = zeros(1,dims);
    for j=1:dims
        dimensionalPositions = zeros(NP,1);
        for i=1:NP
            dimensionalPositions(i) = position(i, j);
```

```
            end
            dimensionalMedians(j)=median(dimensionalPositions);
        end

        averageDistance = zeros(1,dimensions);
        for j=1:dimensions
            distancetoMedian = 0;
            for i=1:NP
                distancetoMedian = distancetoMedian + abs(position(i, j) -
    dimensionalMedians(j));
            end
            averageDistance(j)=distancetoMedian/NP;
        end

        dimensionalDiversity=sum(averageDistance(:));
        averageDiversity = dimensionalDiversity/dimensions;
        if averageDiversity > highestDiversity
            highestDiversity = averageDiversity;
        end
        percentualDiversity = averageDiversity*100/highestDiversity;

        % Terminate the process if the diversity is less than 2 percent
        if percentualDiversity < 2
            break;
        end

        % Optionally, adjust parameters such as inertia weight
        % w = UpdateInertiaWeight(w, t, T);
    end

    % Output the best solution found
    [global_best_fitness, idx] = min(fitness);
    global_best_individual = position(idx, :);
end
```

Index

D

E

F

G

H

I

K

L

M

About the Authors

Mario Alberto Navarro Velázquez obtained a Master of Science in Electronic Engineering and Computer Science in 2019, focusing his research on the design of metaheuristic algorithms and applications in image segmentation. In 2023, he obtained a Ph.D. in Electronic and Computer Engineering at the University Center for Exact Sciences and Engineering (CUCEI) in Guadalajara, Mexico, concentrating on the coevolution of metaheuristic strategies to solve various optimization problems. His most recent recognitions include becoming a part of the National System of Researchers obtaining the distinction as a national researcher level 1 (SNI 1), and membership in the Mexican Academy of Computing (AmexComp). His research interests include artificial intelligence, specifically the design and hybridization of evolutionary algorithms, the development of operators and hyper heuristics to solve high-dimensional problems, and the integration of evolutionary algorithms and machine learning.

Bernardo Morales Castañeda obtained a Bachelor's in Computer Engineering from the University of Guadalajara (UdeG), Mexico in 2016 and a Master's degree in Electronic and Computer Engineering from University of Guadalajara (UDG) in 2018. He obtained a Doctorate degree in Electronics and Computer Science in 2022. Since 2022, Dr. Morales has been serving as a research professor in the Department of Information and Knowledge-based Innovation at the University Center for Exact Sciences and Engineering (CUCEI) of UDG. His research areas include various branches of AI such as artificial neural networks, computer vision, image segmentation, and the development of metaheuristic algorithms. Since 2023, Dr. Morales has been recognized as a member of the National System of Researchers (SNI) with the distinction of National Researcher Level 1.

Itzel Aranguren obtained a degree in Biomedical Engineering from the Universidad de Guadalajara (UDG), Mexico (2016). In 2018, she obtained a Master's of Science in Electronic Engineering and Computing from UDG and successively, and a Doctor in Electronics and Computer Sciences in 2022. Since 2019, Dr. Aranguren has been a research professor in the Department of Computational Sciences of the University Center for Exact Sciences and Engineering (CUCEI) of the UDG. Dr. Aranguren develops her research in medical image enhancement, metaheuristic algorithms, optimization, and vision. She is recognized as a member of the National System of Researchers (SNI), having the distinction of National Researcher Level 1.

Diego Oliva (Senior Member, IEEE) received a B.Eng. in Electronics and Computer Engineering from the Industrial Technical Education Center (CETI) of Guadalajara, Mexico, in 2007, and a M.Sc. in Electronic Engineering and Computer Sciences from the University of Guadalajara, Mexico, in 2010. He obtained a Ph.D. in Informatics in 2015 from the Universidad Complutense de Madrid. Currently, he is a Professor at the University of Guadalajara in Mexico. He has the National Researcher Rank 2 distinction by the Mexican Council of Science and Technology. In 2022, he obtained the distinction of Highly Cited Researcher by Clarivate (WoS). He has been listed in the world's 2% most cited scientists according to Stanford University and Elsevier since 2022. He also serves as editor for journals such as IEEE Access, Engineering Applications of Artificial Intelligence, Swarm and Evolutionary Computation, and Knowledge-Based Systems, among others. His research interests include evolutionary and swarm algorithms, hybridization of evolutionary and swarm algorithms, and computational intelligence.

Marco Perez-Cisneros (Senior Member, IEEE) has a B.Eng. in communications and electronics engineering from the University of Guadalajara, Mexico, a M.Eng. from ITESO University Mexico, and the Ph.D. from The University of Manchester, UK. He is a Professor with the Electro-Photonics Department and has been appointed as the Chancellor of the University Centre of Exact Sciences and Engineering, University of Guadalajara. He is a member of the National Research System in Mexico. Since 2018, he has been a member of the Mexican National Science Academy. He is a Regular Member of the IET, UK. He also serves as an Associate Editor for IEEE Letters.